Advance praise for

GENESIS

"AI may be one of the greatest technological revolutions ever, and the biggest question is how humans will adapt. This important book offers one of the first real looks at the future now in front of us — a future of almost limitless possibility, along with very complex new challenges."

—Sam Altman

"A timely exploration of the relationship between artificial intelligence and knowledge, power, and politics, this book pushes us to think hard about the risk and potential AI holds for humanity."

—Bill Gates

"In the coming Age of Artificial Intelligence, what will be the role of humans? In the final years of his life, Henry Kissinger immersed himself in studying AI, and he coauthored this book with technologists Eric Schmidt and Craig Mundie. It is a profound exploration of how we can protect human dignity and values in an era of autonomous machines."

—Walter Isaacson

"Kissinger, Mundie, and Schmidt provide the deepest reflections we yet have on the opportunities and challenges posed by the looming AI-shaped global system. Readers of their book will learn something profoundly important. Before we can even think about new policies regarding AI, we will need to develop new conceptions of human reason and humanity itself. This book was Henry Kissinger's final work. It may well prove his most prophetic and important. It is profoundly important reading."

—**Larry Summers**

"The next great technological revolution—in artificial intelligence—is already happening. While much of the conversation is about what AI can do and where AI will go, this book brilliantly reframes the discussion. How will human beings relate to AI? How does this thrilling, terrifying new scientific explosion change our conception of what it means to be human? You would expect a profound book given the three authors involved—and you will get it."

—**Fareed Zakaria**

"The authors of *Genesis* raise profound questions that are best answered by placing intelligent tools and technologies in the hands of people, empowering them with real agency to be more confident, more capable, and more in control."

—**Satya Nadella**

"A must-read for anyone trying to think seriously about the challenges posed by AI. *Genesis* captures what we know—and, most important, don't know—about the dangers posed by the unconstrained advance of AI. Drawing on lessons learned in the nuclear age, Kissinger and his colleagues illuminate the murky path ahead."

—**Graham Allison**

"Kissinger, Schmidt, and Mundie have crafted a road map for navigating our near-future, in which unimaginably powerful and ubiquitous AI systems have become autonomous. Their insights into the practical and philosophical implications of humanity's first encounter with a superior intelligence are sobering and inspiring, challenging us to rethink our relationship with technology and our place in the universe. *Genesis* is vital reading for anyone seeking to understand how AI will reshape our world and what it takes to remain human in the age of intelligent machines."

—**Ian Bremmer**

"What does AI mean for discovery? For truth? For security, prosperity and politics? In answering these questions, these three extraordinary thinkers are (characteristically) unafraid to tackle the biggest themes and most profound questions around the dominant technology of our times. Epic in scope, bracing in clarity, and always rooted in deep experience, this is an essential read."

—**Mustafa Suleyman**

"*Genesis* is thought-provoking in the best way—a much-needed exploration of AI's implications for humanity's progress and what makes us human. It is also a road map for how we can harness AI's possibilities, address its challenges, and ultimately coexist with intelligent machines in the age of AI."

—**James Manyika**

"Artificial intelligence boggles the mind, and we struggle to comprehend its promises and perils. In his final book, it is fitting that the master of Grand Strategy, Henry Kissinger, with his superb coauthors Eric Schmidt and Craig Mundie, have focused on this topic. *Genesis* is the book our world needs to read today."

—**Arthur C. Brooks**

"As we try to navigate a responsible path into the future of AI, this book establishes a hopeful framework for how we might coexist while maintaining what it means to be human. In his final work, Henry Kissinger, one of the most consequential thinkers of our time, partnered with Eric Schmidt and Craig Mundie to help guide us into this unprecedented frontier, so that we might harmonize its growth and importance with the wisdom needed to ensure that—this time—technology will be used for the good of humankind. It is a must-read for the decision makers—which is all of us."

—**Condoleezza Rice**

GENESIS

BY HENRY A. KISSINGER

*A World Restored: Metternich, Castlereagh
and the Problems of Peace, 1812–22*

Nuclear Weapons and Foreign Policy

*The Necessity for Choice: Prospects of American Foreign
Policy*

White House Years

Years of Upheaval

Diplomacy

Years of Renewal

*Does America Need a Foreign Policy?
Toward a Diplomacy for the 21st Century*

*Ending the Vietnam War: A History of America's Involvement
in and Extrication from the Vietnam War*

Crisis: The Anatomy of Two Major Foreign Policy Crises

On China

World Order

*The Age of AI: And Our Human Future
(with Daniel P. Huttenlocher and Eric Schmidt)*

Leadership: Six Studies in World Strategy

BY ERIC SCHMIDT

*The New Digital Age: Transforming Nations,
Businesses, and Our Lives*

How Google Works

*Trillion Dollar Coach: The Leadership Playbook of
Silicon Valley's Bill Campbell*

*The Age of AI: And Our Human Future
(with Daniel P. Huttenlocher and Henry Kissinger)*

GENESIS

ARTIFICIAL INTELLIGENCE, HOPE, AND THE HUMAN SPIRIT

HENRY A. KISSINGER

CRAIG MUNDIE

ERIC SCHMIDT

WITH ELEANOR RUNDE
FOREWORD BY NIALL FERGUSON

LITTLE, BROWN AND COMPANY
New York   Boston   London

Little, Brown and Company
Hachette Book Group
1290 Avenue of the Americas, New York, NY 10104
littlebrown.com

First Edition: November 2024

Little, Brown and Company is a division of Hachette Book Group, Inc. The Little, Brown name and logo are trademarks of Hachette Book Group, Inc.

The publisher is not responsible for websites (or their content) that are not owned by the publisher.

The Hachette Speakers Bureau provides a wide range of authors for speaking events. To find out more, go to hachettespeakersbureau.com or email hachettespeakers@hbgusa.com.

Little, Brown and Company books may be purchased in bulk for business, educational, or promotional use. For information, please contact your local bookseller or the Hachette Book Group Special Markets Department at special.markets@hbgusa.com.

Print book interior design by Bart Dawson

ISBN 9780316581295
Library of Congress Control Number: 2024945132

Printing 1, 2024

MRQ

Printed in Canada

To Dr. Kissinger:
statesman, diplomat, mentor, and friend.
We salute you.

CONTENTS

FOREWORD

Niall Ferguson

WHEN HENRY KISSINGER published his essay "How the Enlightenment Ends" in the *Atlantic* in June 2018, many people were surprised that the elder statesman's elder statesman had a view on the subject of artificial intelligence. Kissinger had just turned 95. AI was not yet the hot topic it would become after OpenAI released ChatGPT in late 2022.

As Kissinger's biographer, however, I found it quite natural that the topic of AI gripped his attention. He had, after all, come to public prominence in 1957 with a book about a new and world-changing technology. *Nuclear Weapons and Foreign Policy* was a book so thoroughly researched that it won the approval even of Robert Oppenheimer, who described it as "extraordinarily well informed, and in this respect quite unprecedented in the field of nuclear armament...scrupulous in its regard for fact, and at once passionate and tough in argument."

Although as a doctoral student Kissinger had immersed himself in the diplomatic history of early-nineteenth-century Europe, he was keenly aware throughout his career that the eternal patterns of great-power politics were subject to periodic disruption by technological change. Like so many members of his generation who served in World War II, he had seen for himself not only the mass death and destruction that could be inflicted by modern weapons, but also the dire consequences for his fellow Jews of what Churchill had memorably called the "perverted science" of Hitler's Third Reich.

Contrary to his unwarranted reputation as a warmonger, Kissinger was strongly motivated throughout his adult life by the imperative to avoid World War III—the widely feared consequence if the Cold War between the United States and the Soviet Union turned hot. He understood only too well that the technology of nuclear fission would make another world war an even greater conflagration than World War II. Early in *Nuclear Weapons and Foreign Policy*, Kissinger estimated the destructive effects of a ten-megaton bomb dropped on New York and then extrapolated that an all-out Soviet attack on the fifty largest U.S. cities would kill between 15 and 20 million people and injure between 20 and 25 million. A further 5 to 10 million would die from the effects of radioactive fallout, while perhaps another 7 to 10 million would become sick. Those who survived would face "social disintegration." Even after such an attack, he noted, the United States would still be able to inflict comparable devastation on the Soviet Union. The conclusion was obvious: "Henceforth the only outcome of an all-out war will be that both contenders must lose." There could be no winner in such a conflict, Kissinger argued in his 1957 essay "Strategy and Organization," "because even the weaker side may be able to inflict a degree of destruction which no society can support."

Yet Kissinger's youthful idealism did not make him a pacifist. In *Nuclear Weapons and Foreign Policy*, he was quite explicit that "the horrors of nuclear war [were] not likely to be avoided by a reduction of nuclear armaments" or, for that matter, by systems of weapons inspection. The question was not whether war could be avoided altogether but whether it was "possible to imagine applications of power less catastrophic than all-out thermonuclear war." For if it were not possible, then it would be very hard indeed for the United States and its allies to prevail in the Cold War. "The absence of any generally understood limits to war," Kissinger warned in "Controls, Inspections, and Limited War," an essay published in *The Reporter,* "undermines the psychological framework of resistance to Communist moves. Where war is considered tantamount to national suicide, surrender may appear the lesser of two evils."

It was on this basis that Kissinger advanced his doctrine of limited nuclear war, as laid out in "Strategy and Organization":

> Against the ominous background of thermonuclear devastation, the goal of war can no longer be military victory as we have known it. Rather it should be the attainment of certain specific political conditions which are fully understood by the opponent. The purpose of limited war is to inflict losses or to pose risks for the enemy out of proportion to the objectives under dispute. The more moderate the objective, the less violent the war is likely to be.

This would necessitate understanding the other side's psychology as well as its military capability.

At the time, many people recoiled from Kissinger's seemingly cold-blooded contemplation of a limited nuclear war. Some scholars, such as Thomas Schelling, disputed that an unstoppable escalation could be avoided; even Kissinger himself later distanced himself from his own argument. Yet both superpowers went on to build and deploy battlefield or tactical nuclear weapons, following precisely the logic that Kissinger had outlined in *Nuclear Weapons and Foreign Policy*. Limited nuclear war might not have worked in theory, but military planners on both sides behaved as if it might work in practice. (Indeed, such weapons exist to this day. The Russian government has threatened to use them on more than one occasion since its invasion of Ukraine became bogged down.) The young Kissinger was more right about nuclear weapons than even he knew.

Kissinger never ceased to ponder the implications of technological change in the political realm. In a long-forgotten paper that he wrote for Nelson Rockefeller in January 1968, Kissinger looked ahead to the ways in which computerization might help officials cope with the constantly increasing flow of information generated by U.S. government agencies. As he saw it, senior officials were in grave danger of drowning in data. "The top policy-maker," he wrote, "has so much information at his disposal that in crisis situations he finds it impossible to cope with it." Decision-makers needed to be "consistently briefed on likely trouble spots," Kissinger argued, including potential trouble spots "even when they have not been assigned top priority." They also needed to be furnished with "a set of action-options...outlin[ing] the major alternatives in response to foreseeable circumstances with an evaluation of the probable consequences, domestic and foreign, of each such alternative."

To achieve such comprehensive coverage, Kissinger acknowledged, would require major investments in programming, storage, retrieval, and graphics. Fortunately, the "hardware technology" now existed to perform all four of these functions:

> [W]e can now store several hundred items of information on every individual in the United

States on one 2,400 foot magnetic tape.... [T]hird-generation computers are now capable of performing basic machine operation in nano seconds, i.e., billionths of a second.... [E]xperimental time-sharing systems have now demonstrated that multiple-access capability for large-scale digital computers is possible to allow for information input/output at both the executive and operator stations distributed around the world.... [And] very shortly color cathode ray tube display will be available for computer output.

Later, after his first year in the White House as Richard Nixon's National Security Advisor, Kissinger attempted to obtain such a computer for his own use. The CIA denied the request, presumably because Kissinger without a computer was as much as the intelligence community could handle.

Henry Kissinger never retired. Nor did he ever stop worrying about the future of humanity. Such a man was hardly going to ignore one of the most consequential technological breakthroughs of his later life: the development and deployment of generative artificial intelligence. Indeed, the task of understanding the implications of this nascent technology consumed a significant portion of Kissinger's final years.

Genesis, Kissinger's final book, was co-authored with two eminent technologists, Craig Mundie and Eric Schmidt, and it bears the imprint of those innovators' innate optimism. The authors look forward to the "evolution of *Homo technicus*—a human species that may, in this new age, live in symbiosis with machine technology." AI, they argue, could soon be harnessed "to generate a new baseline of human wealth and well-being... [that] would at least ease if not eliminate the strains of labor, class, and conflict that previously have torn humanity apart." The adoption of AI might even lead to "profound equalizations... across race, gender, nationality, place of birth, and family background."

Nevertheless, the eldest author's contribution is detectable in the series of warnings that are the book's leitmotif. "The advent of artificial intelligence is," the authors observe, "a question of human survival.... An improperly controlled AI... could accumulate knowledge destructively...." Here, rephrased for *Genesis* but immediately recognizable, is Kissinger's original question from his 2018 *Atlantic* essay "How the Enlightenment Ends":

> [AI's] objective capacity to reach new and accurate conclusions about our world by inhuman methods not only disrupts our reliance on the scientific method as it has been pursued continuously for five centuries but also challenges the human claim to an exclusive or unique grasp of reality. What can this mean? Will the age of AI not only fail to propel humanity forward but instead catalyze a return to a premodern acceptance of unexplained

authority? In short: are we, might we be, on the precipice of a great reversal in human cognition — a dark enlightenment?

In what struck this reader as the book's most powerful section, the authors contemplate a deeply troubling AI arms race. "If...each human society wishes to maximize its unilateral position," the authors write, "then the conditions would be set for a psychological contest between rival military forces and intelligence agencies, the likes of which humanity has never faced before. Today, in the years, months, weeks, and days leading up to the arrival of the first superintelligence, a security dilemma of existential nature awaits."

If we are already witnessing "a competition to reach a single, perfect, unquestionably dominant intelligence," then what are the likely outcomes? The authors envision six scenarios, by my count, none of them enticing:

1. Humanity will lose control of an existential race between multiple actors trapped in a security dilemma.

2. Humanity will suffer the exercise of supreme hegemony by a victor unharnessed by the checks and balances traditionally needed to guarantee a minimum of security for others.

3. There will not be just one supreme AI but rather multiple instantiations of superior intelligence in the world.

4. The companies that own and develop AI may accrue totalizing social, economic, military, and political power.

5. AI might find the greatest relevance and most widespread and durable expression not in national structures but in religious ones.

6. Uncontrolled, open-source diffusion of the new technology could give rise to smaller gangs or tribes with substandard but still substantial AI capacity.

Kissinger was deeply concerned about scenarios such as these, and his effort to avoid them did not end with the writing of this book. It is no secret that the final effort of his life—which sapped his remaining strength in the months after his hundredth birthday—was to initiate a process of AI arms limitation talks between the United States and China, precisely in the hope of averting such dystopian outcomes.

The conclusion of *Genesis* is unmistakably Kissingerian:

What some see as an anchor to steady ourselves in the storm, others see as a leash holding us back. What some praise as necessary steps toward a pinnacle of human potential, others see as a headlong rush into an abyss.

In this case, instinctive emotional divergences—and the subjective lines that are drawn by all parties—will create an unpredictable and combustible situation. Increasingly stark positions of potential "winners" and "losers" will intensify the pressure of these circumstances. The fearful will slow their own development and sabotage that of others. The overconfident will disguise their powers and, in secret, speed up their work. The timeline of coming crises will be accelerated beyond prior human experience; quickly, we will be engulfed, and it is not clear whether or how we will survive.

The technologist's habitual response to such forebodings is to remind us of the tangible benefits of AI, which are already very obvious in the realm of medical science. I do not disagree. In my view, AlphaFold—a neural-network-based model that predicts three-dimensional protein structures—was a far more important breakthrough than ChatGPT. Yet medical science made comparable advances in the twentieth century. The world wars and the Holocaust nevertheless occurred, even as antibiotics, new vaccines, and countless other therapeutics were discovered and made widely available.

The central problem of technological progress manifested itself in Henry Kissinger's lifetime. Nuclear fission was discovered in Berlin by two German chemists, Otto Hahn and Fritz Strassmann, in 1938. It was explained theoretically (and named) by the Austrian-born physicists Lise Meitner and her nephew Otto Robert Frisch in 1939. The possibility of a nuclear chain reaction leading to "large-scale production of energy and radioactive elements, unfortunately also perhaps to atomic bombs" was the insight of the Hungarian physicist Leó Szilárd. The possibility that such a chain reaction might also be harnessed in a nuclear reactor to generate heat was also recognized at that time. Yet it took little more than five years to build the first atomic bomb, whereas it was not until 1951 that the first nuclear power station was opened.

Ask yourself: Which did human beings build more of in the past eighty years: nuclear warheads or nuclear power stations? Today there are approximately 12,500 nuclear warheads in the world, and the number is currently rising as China adds rapidly to its nuclear arsenal. By contrast, there are 436 nuclear reactors in operation. In absolute terms nuclear electricity generation peaked in 2006, with the share of total world electricity production that is nuclear declining from 15.5% in 1996 to 8.6% in 2022, partly as a result of political overreactions to a small number of nuclear accidents whose impacts on human health and the environment were negligible compared to the effects of carbon dioxide emissions from fossil fuels.

The lesson of Henry Kissinger's lifetime is clear. Technological advances can have both benign and malign consequences, depending on how we collectively decide to exploit them. Artificial intelligence is of course different from nuclear fission in a host of ways. But it would be a grave error to assume that we shall use this new technology more for productive than for potentially destructive purposes.

It was this kind of insight, born of historical as well as personal experience, that inspired Henry Kissinger to devote so much of his life to the study of world order, and the avoidance of world war. It was what made him react with such alacrity — and concern — to the recent breakthroughs in artificial intelligence. And it is why this posthumous publication is as important as anything he wrote in the course of his long and consequential life.

OXFORD
JULY 2024

IN MEMORIAM: HENRY A. KISSINGER

ON NOVEMBER 29, 2023, Dr. Henry A. Kissinger passed away at the age of one hundred. An inspiration to all who knew him, he worked to the very end on this volume, his twenty-second book. During our frequent meetings in his final year of life, he would firmly insist on the importance of our subject and on the urgent need to broadcast its message. We, as his two coauthors, were among the last people to see and speak with him in the days before his death. In now concluding this project on his behalf and at his request, we have strived to preserve the originality of his thought and the ringing tenor of his voice on a matter of the utmost importance to the future of humanity. To finish what he started, so that this final literary undertaking does not die with him but lives on in the world without him, is but a small contribution in his memory.

After having done so much to build our world, he spent his final hours in the vital effort to save it. Indeed, his final written word is a request for all humanity to continue the vast project of securing the future of our species. In the middle years of the previous century, Dr. Kissinger was a chief architect in the philosophical and diplomatic effort to shield humanity from atomic annihilation—the twentieth century's encounter with the grim realities of existential risk. This courageous defender against that risk departs just as a new risk now arrives. His life ends just as a new form of life begins. As we face the dawn of the age of AI, we are cognizant of how few besides Dr. Kissinger—student of the nineteenth century, master of the twentieth, and oracle of the twenty-first—have been as well-positioned as he to lay the groundwork for our future.

Dr. Kissinger was first and foremost a philosopher of history. Emerging from his deep investigations on the subject of tragedy was a lifelong quest to demonstrate that idealism of heart could be compatible with, and ennobling of, realism of mind. One could hold, in the formulation of the French writer Romain Rolland, both the "pessimism of intellect" and the "optimism of will." Where the optimist aspires to assured human control of our affairs, the pessimist sees our condition as determined by forces beyond our control: the laws of nature and the cycles of history.

Certainly, he knew all too well how a fervent idealism can be taken advantage of by ideologues without remorse in the spilling of blood or hesitancy in the enforcement of might. Fascism, Communism, totalitarianism, militant religious fanaticism—all have laid claim to the most idealistic ends pursued in our history. First a victim of, then a military and diplomatic combatant against, such inhuman excesses, he undertook to help rebuild the world anew, atop a foundation of order without shame and safety without guilt. Through his active management of international affairs, Dr. Kissinger steered his adopted country—and the world—through uncertain upheavals, grounding himself in the hard soils of historical fact and national interest.

Brilliant as he was about the selective necessity of realism, Dr. Kissinger was also an idealist—respecting, as his biographer Niall Ferguson puts it, "the role of human freedom, choice, and agency in shaping the world." In theory and through practice, he demonstrated his belief that humans do not, and cannot, live as if the future is inevitable. In his senior thesis at Harvard, *The Meaning of History*, 27-year-old Henry Kissinger grappled with the same philosophical debate that now animates his last work: "Whatever one's conception about the necessity of events, [and]...however we may explain actions in retrospect, their accomplishment occurred with the inner conviction of choice."

To him it was not certain that humanity would survive inhuman designs crafted from the fires of our own forge. Faced with—and burdened by—the daunting prospect of nuclear catastrophe, he succumbed neither to the fatalism of determinism nor to prophecies of doom. Granted, existential fears could give rise to nihilism, but they could also infuse the best among us with the needed forcefulness to defy evil and to defend what would need to be preserved for the future of our species. In the early 1950s, as a young professor at Harvard, he participated in a set of meetings at which leading scientists and academics like himself gathered to discuss and debate the potential consequences of, and the measures needed in order to prevent, nuclear warfare. From those meetings, doctrines emerged that have since kept our world safe from the participants' worst fears.

Decades later, conversing with the two of us, he would speak often about those meetings—their structure, their purpose, and their retrospective importance. His view until the end remained the same, at once unresigned to the march of fate and unsubscribed to any vision of utopia. The same balance obtains in AI as in the nuclear context: Small groups of dedicated individuals can alter history by stepping in and manifesting their values. At the same time, however, and no matter the genius of the scientists who were building new intelligences, he believed that their training would not in itself suffice to ensure the necessary minimum of safety and security in the operation of these newest instruments.

To that end, his legacy in AI is not limited to purely philosophical and scholarly explorations but also encompasses practical proposals. A half century after his secret flight to Beijing and the subsequent opening of relations between the United States of America and the People's Republic of China, Dr. Kissinger took one more trip to the Chinese capital. He went at the urgent and specific invitation of President Xi Jinping to discuss, as the principal topic, the risks humanity faces with AI. It was the last foreign trip he would ever make and his final diplomatic mission.

If, in the past, Dr. Kissinger elevated to an art the study and practice of statecraft, today his search for answers has elevated AI to a matter of more than science. With one of us, and with Professor Daniel Huttenlocher of MIT, he authored *The Age of AI: And Our Human Future*, published in 2021, which predicted that the arrival of artificial intelligence would create a new epoch in history, similar in its impact to the eighteenth-century Enlightenment for its ability to change human thought in profound and unexpected ways. In this new age, however, rather than working forward from questions posed by humans, humanity confronts answers provided by AI to questions that no human ever asked. As AI conquers the realms of human knowledge, Dr. Kissinger sought to rely and to draw upon the resources of human wisdom.

In this volume, we explore with Dr. Kissinger the impact of AI on eight different areas of human activity and thought, culminating with his own philosophical answer to the ongoing search for a viable strategy to balance benefits and risks. In pursuit of that goal, he explores the prospects of a human coexistence with AI and, in due time, of human-AI coevolution. In conceptually unlocking the possibility of a reconciliation of these two species—the one organic, the other synthetic—he also reveals the need for a choice: to create a world in which *AI* becomes more like *us*, or one in which *we* become more like *AI*.

Since the publication of his first book on the age of AI, Dr. Kissinger increasingly came to perceive a limit to the utility of reason in the final days of such an age. To us humans, explanations beyond our own understanding—or our own making—can appear utterly baffling; our instinct is to presume that they are less advanced and more primitive than our scientific explanations, a step backward and not forward. But that is a dangerous assumption.

If, in the words of Arthur C. Clarke, "Any sufficiently advanced technology is indistinguishable from magic," and if miracles are engineered from mathematics, the future should be inexplicable, bewildering, even magical. Over the collective decades that we have known Dr. Kissinger, he has generously taught us from his learnings of statecraft—that complex realm of human affairs—that although reason has been the dominant paradigm through which humans have mastered our world, it cannot be the paradigm with which we master ourselves.

Moving forward, we should therefore not expect to rely solely on reason—the historical fuel for our greatest human advancements. But neither can we fully abandon it. Not unlike his moderation between idealism and realism, his final investigations into our future achieve an equilibrium between that empirical quality of truth and something else—philosophically beyond reason but chronologically lagging behind. Just as foreign policymaking cannot afford an excess of either extreme, neither can our framework for the future.

AI, thus, is a unique challenge that requires thinking that might seem, at first, irrational or overwrought—and indeed, the scenarios described in this book are startling. But in counseling us and others that he was himself but a humble student both of humanity and of its latest and potentially final creation, he impressed upon us that the greatest danger posed by AI would be for us to declare too early, or too completely, that we understand it.

His depth of intellect and perception of people are not qualities we expect to encounter again. We know of no one else who, at the age of 93, could master an entirely new and previously unknown field of technical knowledge. With his unquenchable curiosity and mental vitality, coupled with a devotion to work and sense of mission, no pain of body or spirit was ever enough to dampen his passion for progress. Regardless of the infirmities of old age, he rose each day with undefeatable resolve to move the world forward. His indomitable strength came, perhaps, from his unmatched discipline—hardened in youth by oppression, sculpted by service in war, and tested for decades by the strife of public life.

IN MEMORIAM: HENRY A. KISSINGER

We are but two of many whose lives have been shaped by this extraordinary man. We will miss him dearly—no doubt, in more ways than we can now imagine. Departing us on the eve of great uncertainty, he is needed now more than ever. That is why, for this book, there seemed no title more apt than *Genesis*—a new beginning for him as for all humans. Whether humanity succeeds or fails, he will no longer be present to witness the ultimate outcome of his efforts. At least now we have his wisdom to guide our own.

—Eric Schmidt and Craig Mundie

GENESIS

INTRODUCTION

A FEW SHORT YEARS ago, artificial intelligence (AI) inhabited a small corner of the public debate. Today, following rapid advances, AI is a front-page topic at news outlets everywhere and an issue on the minds of leaders in science, business, journalism, public service, education, and politics the world over.

As we see it, both the general public and many experts in the field continue to overlook important aspects of this new age of AI. New forms of AI and human responses to them could transform nothing less than the human relationship to reality and to truth, the exploration of knowledge as well as the physical evolution of humanity, the conduct of diplomacy, and the international system. These are among the crucial issues of the coming decades, and they ought to be the guiding concerns of leaders in every arena.

The latest capabilities of AI, impressive as they are, will appear weak in hindsight as its powers increase at an accelerating rate. Powers we have not yet imagined are set to infuse our daily lives. Future systems will facilitate enormous and largely beneficial advances, improving our health while generating wealth.

But these capabilities come with technical and human risks, some of which are known and some unknown. Today's technologies already function in ways that their inventors did not predict, and this pattern is likely to continue. Each fruitful research path taken by our scientists—and there will be more than one—could yield new branches of unforeseen powers that may or may not be comprehensible or beneficial to humans.

AI seems to compress human timescales. Objects in the future are closer than they appear. As just one example, machines with the ability to define their own objectives are not far away. If we have any hope of keeping up with the risks involved, we must respond and act within the shortest conceivable timeline. Cognizant of the stakes and the urgency of the task ahead, we note here only a few of its many facets.

As human-machine partnerships become ubiquitous, humans will have to determine these relationships' proper nature. The answers may be taken from the logic of security and efficiency, derived via the study of history, or discerned from the divine. Individuals, nations, cultures, and faiths will need to determine the limits, if any, of AI's authority over truth. They will need to decide whether to allow AI to become an intermediary between humans and reality. In this connection, they will need to choose between, on the one hand, retaining the traditional role of human enterprise (while likely ceding leadership to AI in the discovery of new knowledge) and, on the other hand, abandoning the biologically limited human mind in favor of a potentially reengineered partnership with AI on the intellectual frontier. Do we pick our objectives and harness AI to achieve them, or do we let AIs help pick some of the objectives themselves? Most urgently, humanity must give human dignity a modern and sustainable definition that can provide a philosophical orientation for our decisions in the years to come.

The advent of artificial intelligence is, in our view, a question of human survival. As we explain later in this book, AI's future faculties, operating at inhuman speeds, will render traditional regulation useless. We will need a fundamentally new form of control. For the global scientific community, the immediate task is to find technical measures for instilling intrinsic safeguards in every AI system. For their part, nations and international organizations, once they have coalesced around a consensus, must develop new political structures for monitoring, enforcement, and crisis response. That will require the resolution of not one but two "alignment problems": the technical alignment of human values and intentions with the actions of AI, and the diplomatic alignment of humans with their fellow humans.

Dr. Henry Kissinger, one of this book's coauthors, closely mentored his two collaborators on the latter problem, while they—as previous senior leaders of Microsoft and Google—coached him on the former. Craig Mundie was Microsoft's chief technology-policy liaison to Washington and foreign governments worldwide, concurrently overseeing Microsoft Research before more recently advising the leadership of the research organization OpenAI. Eric Schmidt led Google for a decade before spending the next decade forging a network of talent and ideas at the intersection of technology, philanthropy, and security with the aim of safeguarding humanity.

The urgency of the issues that we have confronted together is such that, rather than waiting for a crisis, we deem it imperative that our society, and indeed our species, address them proactively. And, although the cause of human safety is one necessary component of a successful response to AI, it cannot answer all the questions that AI raises—for, in the age of AI, humanity will change. The question is whether and to what extent we humans will choose to continue to assert authority over how that change occurs.

PART I

IN THE BEGINNING

CHAPTER 1

DISCOVERY

D ISCOVERY MAY BE the single most exhilarating capacity of the human species. Driven by curiosity, delighted by surprise, we fill the vacuums we perceive and transform the questions we ask into answers. Exploration is so integral to our self-definition that despite manifest dangers and frustrations, we have continued, unrelenting, down its many paths.

Throughout history, human exploration of, in particular, our physical environment has been a story of immense courage in the face of severe risk. Human teams setting out on such endeavors were often in contest with their own mortality. In the early sixteenth century, Ferdinand Magellan's circumnavigation of the globe was a three-year odyssey filled with violence, starvation, and death. Magellan's voyage was the first successfully to steer a course around the globe. In the process, he smashed the endurance record on the high seas, established the dimensions of the planet, and—taking place as the enterprise did in the context of European colonialism—opened the way for social and economic exchange on an international scale.

Most of Magellan's sailors recognized they were risking catastrophe. Though it was no longer widely believed that the Earth was flat, its roundness had not been proved as fact, and many in his crew may have feared sailing off the edge of the world.

Both he and his crew were aware they would pay heavily for any mistaken assumptions. And indeed they did err, and they did pay: They had underestimated the length of their journey, the quantities of food needed to sustain them, the dangers of mass poisonings, and the risk of damaged or otherwise inoperative ships. Of the original fleet of five vessels carrying approximately 270 crewmembers, only one solitary ship bearing eighteen ghostly survivors returned to port in Spain. Not among them was their captain, who had perished along the way after being hit in the leg by a poisoned arrow.[1]

In the 400 years after Magellan's time, every corner

of the world would be mapped—except Antarctica, a land as desolate as an alien planet. The Anglo-Irish explorer Ernest Shackleton would come closer than anyone before him to reaching that land's South Pole: the bottom of the world. In 1909, with an inexperienced crew and with support from no government but only a few private loans and individual contributions, Shackleton and his men set the record for the longest such journey and paved the way for explorers after them.

Although unable to claim he was the first to reach the South Pole itself, Shackleton nevertheless earned the respect of later generations. This he did by prioritizing human values over exploratory ambitions. One year into the expedition, the team members, harnessed daily to sledges for brutal ten-hour shifts, were advancing only a few miles each day. Although they had enough food to reach the Pole, the remaining supply would not suffice to see them back to their ship.

Then and there, within 97 miles of triumph, Shackleton, rather than risking the lives of his men, made the decision to turn back. "We have done our best," he declared in his journal.[2] On their retreat, Shackleton offered his one daily biscuit to Frank Wild, an ailing crew member, who would write in his diary that "All the money that was ever minted would not have bought that biscuit, and the remembrance of that sacrifice will never leave me."[3]

Undeterred by failure, Shackleton would later make further expeditions to the South Pole. For decades, a story (later debunked) circulated that he had placed the following ad in the London *Times*:

> Men Wanted: For hazardous journey. Small wages, bitter cold, long months of complete darkness, constant danger, safe return doubtful. Honour and recognition in case of success.[4]

The ad may have been apocryphal, but the sacrifice it illustrates was not. These were the realities of exploration only a century ago: Our progress on the frontier was limited by the number of brave souls who would accept such long odds.

Perhaps in recognition of these hazards, some governments saw fit to sponsor and reward expeditions of discovery, and these ventures became part of the game of international competition. Magellan's journey, for instance, was defined by politics. Unable to secure financial support from the king of Portugal—his own sovereign—Magellan had defected and sailed instead under the more forthcoming sponsorship of the Spanish crown. After his death, the crew chose a Spaniard, Juan Sebastián Elcano, to assume command. On the return voyage, desperate and virtually bereft of food and supplies, Elcano attempted a stop in the Cape Verde islands—a Portuguese colony off the coast of West Africa—sending thirteen crew ashore to negotiate with the colonial governor. But their request was met with humiliating rejection.

With a broken heart but now even stronger resolve to demonstrate Portugal's folly and Spain's achievement, Elcano ordered the anchor raised and continued onward home. (He left the members of his advance mission behind in unfriendly territory.) Finally completing the journey—and fulfilling Magellan's vision—Elcano wrote to the Spanish king, by now the Holy Roman Emperor and the world's most powerful sovereign:

> Your Majesty will know better than anyone that what we ought most to value and hold on to is that we have discovered and sailed the whole roundness of the world, that going to the West, we have returned from the East.[5]

Eventually, dauntless humans and their political sponsors would push the project of discovery altogether beyond our earthly sphere. We would undertake not just to study the planets—already an ongoing exercise for millennia—but to feel their surfaces beneath our feet. Following the Second World War, the "space race" between the United States and the Soviet Union, a complement of the geopolitical Cold War, brought two superpowers into a competition to send humans where they had never ventured before.

The astronauts themselves could be considered pioneers, but none of the ventures into space undertaken by Moscow and Washington were the product of an individualistic gamble of life and fortune. Rather, each was a coordinated diplomatic and military mission, powered by enormous investments of money, time, and latitude for experimentation. In the past seventy years, more than 600 astronauts have pierced the heavens and toured space, some of whom have gone farther to circle, orbit, or walk on the moon.[6] Thus did U.S.-Soviet rivalry bring us at once to the brink of nuclear annihilation and upward to the stars above.

More than a century before the Age of Discovery—as the early modern era of Magellan, Vasco da Gama, Amerigo Vespucci, and so many others came to be called—Chinese ambitions on the high seas had been unmatched for their scope and scale. Dwarfing the resources and support provided to explorers by governments in the West, the "treasure fleet" of Zheng He, the great Ming-era admiral, comprised dozens, and sometimes even hundreds, of state-of-the-art ships carrying tens of thousands of sailors, soldiers, diplomats, and merchants. Each of the Chinese admiral's ventures took two years. After navigating the Pacific waters near China's southeastern shores, his ships proceeded westward to the Bay of Bengal, the Indian Ocean, the Arabian Sea, the Red Sea, and finally the Swahili coast of East Africa, in explorations stretching through nearly three decades from 1405 to 1433.[7]

Zheng He's voyages, closer in their origins and motives to the modern American and Soviet space programs than to Western equivalents at his time, were not so much leaps into the unknown as the rich products of imperial statecraft. But therein also lies a problem. The very things needed to ensure an expedition's success, even within a well-resourced state, can doom its cause in the long run. Politics change. Priorities shift. Human patience is easily exhausted. The expenditures lavished on the "treasure fleet" were so great that factions within the Ming court began to castigate the emperor's funding of these missions. Political setbacks and natural disasters exacerbated the unrest. In the end, the Chinese government decided to destroy or intentionally neglect its best ships, along with many records of Zheng He's voyages, lest a similar visionary arise again to captivate and seduce the country's leadership. The ships rotted away, and the likes of those vessels were not seen again for 400 years.

Analogously, after the U.S. won the space race, without a competitor to motivate its national efforts, support in Washington for space exploration languished and the budget of the National Aeronautics and Space Administration (NASA) was cut. Over the span of five decades, America's capacity for crewed spaceflight steadily degraded. From being the sole nation on Earth capable of landing humans on the moon, the U.S. fell to having the wherewithal only to transport humans to "low Earth orbit," and then, finally, lost the ability to take anyone to orbit at all.

America's reputation was rescued only by the efforts of private explorers who, led by the aerospace company SpaceX, revived cosmic ambitions in the West. That company's efforts have by now led some to the hitherto unimaginable point of viewing our closest celestial neighbors not as temporary destinations, but as permanent homes. A century ago, Shackleton helped establish a human presence on Earth's southernmost pole. Today, Shackleton Crater—a depression, so named in his honor, at the south pole not of Antarctica but of the moon—is the planned site of humankind's next outpost.

The existence of alternative backers has been critical to sustained exploration. In Magellan's Europe, if one monarch proved unsupportive, an explorer could raise financing from another. By the twentieth century, Ernest Shackleton, aspiring to claim the South Pole for a dwindling British empire distracted by the 1914 outbreak of World War I and unable to raise the necessary funds from the Crown, would turn largely to private donors. The rise of private, for-profit corporations—which allow the pooling of investment and risk—unlocked additional possibilities. One wonders what might have become of Zheng He and his successors if they had had such options.

ENTER AI

For the longest period in Western history, the exploration of reality was focused on geographical entities—our planet and our closest celestial neighbors. As humans steadily acquired mastery over our immediate physical environment—on land, at sea, and in the heavens above—it was only a matter of time before our restless human instinct for discovery would expand its range from the space around us to the ideas within us. Today, we stand at the frontier not of physical but of intellectual exploration.

The development of artificial intelligence has ushered in a new era of discovery. Where AI is integrated into physical systems, robotic sensors assume roles that would earlier have been performed by humans, thereby untethering human discovery from physical danger to the explorer and thus multiplying the ranks of eager developers and investors.

AI also does not experience fear, and so it is undaunted by reality's vast expanse. Nor does it experience shame, and so it unhesitatingly fails—but so quickly can an AI recalibrate that, through constant improvisation and experimentation, it can accommodate high failure rates without causing setbacks to its aforesaid developers and investors.

Today, AI discovery is a project led almost exclusively by private corporations and entrepreneurs, with states emerging as supplementary backers. Even without complementary government action, however, AI growth and expansion are likely to remain fueled by an abundance of diverse sources of capital. True, in today's still-early phases of development, significant amounts of human capital and societal support may continue to be needed; but sustaining AI exploration in the future may cease to be a fiscal and political drain on the societies that deploy it. Absent unforeseen developments and unlike in earlier ages of exploration—all of which ended before they had achieved their full potential—we can expect discovery of and with AI to continue unabated.

Nevertheless, though AI is partially unshackled from earlier constraints on discovery, it cannot escape them all—particularly as its effects become more pronounced. The tolerance for risk among democratic societies and the uncertain future of international gamesmanship will continue to be a significant X factor in the realm of artificial intelligence. Perhaps the story will be one of an "AI race." Perhaps it will result in the equivalent of the Ming government's wholesale destruction of Zheng He's "treasure fleet." Or, perhaps, states will steer progress along some middle course.

THE POLYMATHIC MIND

In hindsight, it seems obvious that the realm of discovery was bound to expand beyond the physical—the stomping ground of admirals, astronauts, and adventurers—and that its explorers would become more various. And indeed, from fairly early on, history would witness the eventual rise of a new—or, if not new, then decidedly different—kind of human discoverer: the polymath.

Exceptional for their ability to master many spheres of knowledge, any of which might ordinarily absorb a lifetime's labor, polymathic individuals throughout history may number, at a guess, in the mere hundreds. Whether devoted to the arts or to the sciences, or both, all have been imbued with a passion to revolutionize, or to create from scratch, entire fields of study. Propelled not so much by the courage of the heart as by the sheer power of the mind, they have ventured undaunted into the depths of human knowledge and human imagination: an even vaster terrain than the one confronting a typical explorer of the physical world.

At times, these singular individuals with their wondrous capacities to decode our universe have been viewed with both awe and suspicion as sorcerers, or as putative intermediaries with the divine Creator of the universe—a reputation that could frequently bring them into conflict with religious or political authorities. At other times, treasured for their gifts, they have been encouraged to pursue their studies under the direct auspices of just such authorities, and have found themselves rewarded for their efforts.

In the Islamic Golden Age, polymaths sought ways to pioneer science in service of faith. Ibn al-Haytham, from Basra in modern-day Iraq, proposed the concept of the scientific method—five centuries before Renaissance contemporaries would lay claim to it in the West.[8] Equally at home in geometry, astronomy, optics, and experiential psychology, al-Haytham also had a deep knowledge of hydraulic engineering.

This last would bring him into conflict with religion. Having laid claim to an ability to regulate the flooding of the Nile, a natural phenomenon then still believed to be of supernatural doing, he was invited to meet the caliph in Baghdad. There his proposed engineering projects were found to be at odds with Islamic theology. As punishment for his audacious claims and revolutionary thinking, he was forced into hiding until the caliph's death.

Other polymaths like Muhammad ibn Musa al-Khwarizmi—a Persian from the territory of today's Turkmenistan—found greater success with areas of scholarship in explicit service of their theological masters. Al-Khwarizmi was appointed as court astronomer and head of the library of the House of Wisdom in Baghdad.[9] Astronomy flourished under the Abbasid caliphs, who became lavish patrons of men like him for their direct contributions to the Islamic faith. For example, the geographical coordinates of sacred locations, and most important the direction of Mecca—knowledge of which is a requirement for purposes of Islamic prayer—had been made much more precisely calculable thanks to medieval astronomy's improved access to the exact positioning of the stars.

Contemporaries of al-Haytham and al-Khwarizmi would find ever more creative ways to sustain the spirit of discovery by deepening the alliance between science and religion, ensuring refuge—for those who possessed, in the words of Ibn Rushd (also known as Averroës), another great Muslim polymath of the same era, "the unity of intellect"—in an age named not for its achievements of reason, as the European Enlightenment was, but for its fervent religiosity.[10]

Thousands of kilometers from Baghdad—and from each other—Chinese and Indian polymaths aligned themselves not with divine authority but with structures of government, thereby sharing a distinctive proximity to politics achieved by both privilege and effort. In the twelfth century, Hemachandra—the "knower of all knowledge in his times"—served as an adviser to King Kumarapala of what is now Gujarat. Centuries later, Akbar the Great—the young Mughal emperor—would both rule the state himself and achieve scholarly successes in architecture, engineering, and literature.

As for Chinese prodigals, in addition to their prowess in intellectual matters, they, too, were deeply engaged in service at court as advisers, scholar-officials, and high-ranking administrators. Government was as much their client as their patron. At any one time of day, these thinkers might be found supervising the formulation of calendrical science in their duties as head officials of the Bureau of Astronomy; an hour later, they could be advising the emperor's central cabinet on how best to increase crop yields. They were commissioned to build grand machines of war, dispatched as ambassadors on diplomatic missions to neighboring kingdoms, and tasked with advising the emperor on matters of economic policy.

But they had no autonomy, being tolerated only to the extent that the emperor sought their intellectual services. In the Chinese world, factional politics became as much a hindrance to genius as clerical restrictions were in the Islamic world. Once-in-a-generation intellects were at the mercy of the very system that had identified their talents in the first place, with science still subservient to the son of heaven. Shen Kuo—a polymath of the Song dynasty—was ousted by a jealous military officer, ultimately falling out of favor with the emperor and driven into isolation as the outcome of a political rivalry with—strangely enough—the dynasty's only other major polymath.[11]

Lone polymaths flourished during ancient and medieval times in the Middle East, India, and China. But it was not until after the Age of Exploration that systematic conceptual investigation commenced first in Europe and later in the United States, in what we now call the Age of Reason or the Age of Enlightenment. Preceded and facilitated by the fifteenth- and sixteenth-century Renaissance, the ensuing four centuries—bringing us up to the beginnings of our AI age—proved to be a fundamentally different era for intellectual discovery.

In pre-Enlightenment times, polymaths had little choice but to serve a higher power, whether that be an emperor or a caliph. By contrast, many leading figures of the European Enlightenment were enabled to pursue their insights not as means to political or theological ends but as ends in themselves. "A man can do all things if he will," the Italian polymath Leon Battista Alberti proudly claimed of the Renaissance Man.

Still, although intelligence is necessary for exploration, it is not sufficient. Beyond maintaining an appetite for risk, explorers must also be supported by the right resources, the right environment, and the right collaborators. During the Enlightenment, explorers could gain access to all three. Governments and corporations, largely owing to their interest in translating scientific theory into military and commercial applications, remained active patrons and partners to Euro-Atlantic polymaths while, for the most part, freeing the latter to direct their energies and skills as they saw fit. Even when attempts were made to co-opt, suppress, or otherwise intervene, Europe was fragmented enough to allow thinkers of ideas unwelcome in one locale to find a home in another. Thus, the Frenchman François-Marie Arouet, better known by the pen name Voltaire, would spend a significant amount of time outside France, while the Russian Mikhail Lomonosov, having determined at age nineteen to "study sciences," would walk from the constraining circumstances of his native village in the far north all the way to Moscow, where he acquired a basic education before studying in Kiev and then in Germany at the University of Marburg and then Freiberg.[12]

One salient result was that human progress was now propelled by a newfound concentration and linkage of like-minded thinkers in physical or mental proximity, forcing intelligence of the highest quality into competition and collaboration. Previously, the story of human prodigy had been a lonely one, siloed by the tyrannies of space and time. Often working in isolation, with rarely anyone else in one's spatial or temporal vicinity with whom to collude and commiserate, polymaths could push the boundary only as far as their own capacities would allow. Moreover, this limited connection to the few other intellectual pioneers within and between polities all but ensured a redundancy of effort by inventors lacking direct knowledge of each other's research and unable to build upon each other's work.

Gradually, for those lucky enough to be in possession of accurate, timely, and faithfully translated material, complex inventions could be pieced together by group effort—not just simultaneously but across generations. By the time of the Enlightenment, polymaths were able to bridge not only disciplines but also separate areas of understanding that had never before been reconciled or amalgamated into a coherent whole. No longer was there Persian science or Chinese science; there was just science.

This capability to integrate knowledge from diverse domains helped to produce rapid polymathic break-throughs, and in time would prove to be the best attempt yet at "collective intelligence." In the twentieth century, for example, the World War II Manhattan Project saw just such a disproportionate density of mental ability as, working together, the era's most brilliant minds translated generations of theoretical physics into devastating appli-cation in under five years—a feat that would have been inconceivable to their predecessors. Institutions such as the Institute for Advanced Study at Princeton Univer-sity and the RAND Corporation in California similarly became sanctuaries for gifted minds.

Naturally, some polymaths, burdened by brilliance, still preferred to work alone. One of them was the Serbian American inventor Nikola Tesla:

> The mind is sharper and keener in seclusion and uninterrupted solitude. No big laboratory is needed in which to think. Originality thrives in seclusion free of outside influences beating upon us to cripple the creative mind. Be alone, that is the secret of invention; be alone, that is when ideas are born. That is why many of the earthly miracles have had their genesis in humble surroundings.[13]

But Tesla was the exception, not the norm. The twentieth century produced a Cambrian explosion of applied science, hurling humanity forward at a speed and scale incomparably beyond previous evolutions. Powered by the combination of their mental capacities, groups of polymaths were also now equipped with the tools of modernity. The aggregate effect of such intense momentum has helped us overcome no small number of human limitations. Digital communication and internet search, themselves the product of polymathic groupings, have enabled the enlargement of such groups and an assembly of knowledge well beyond prior human faculties.

To be sure, there has also been a limit. No matter how well we optimize the design of the vessels intended to carry us to new frontiers, and no matter how well we organize genius into working operations, biological restrictions and human shortcomings continue to constrain our capacities. Our time on Earth is finite. We need sleep. We easily tire. We require rest and respite. Even when on the job, most humans can concentrate on only one task at a time.

Consider, in the late nineteenth and early twentieth century, Thomas Edison—himself a polymath and Tesla's rival—whose thousands of experiments to create a "practical incandescent electric lamp" took three years to achieve success. True, Edison had been partly distracted by contributing improvements to the telephone, the invention of Alexander Graham Bell. But even with the help of many assistants, his quest for a light bulb required extraordinary effort. Still today, the research and development of many cutting-edge technologies remains expensive, all-consuming, long-drawn-out, and both politically and psychologically daunting to invest in. Because the outcomes are so uncertain, probing the remaining physical frontiers—space, the deep sea, Earth's inner crust—is still the project of only the most successful companies and the wealthiest governments.

Of course, it is precisely these same unfavorable odds that make discovery still feel meaningful, even miraculous.

The physicist John von Neumann, who would later be regarded as among the last great polymaths, was named "Person of the Century" by the *Financial Times* for embodying the twentieth century's characteristic confidence in the power of the mind to "harness and tame the physical world."[14] Indeed, von Neumann had devoted himself with ferocious energy to such central issues as mathematical theory and the atomic bomb, but most crucially to the creation of the computer: both the century's last major breakthrough and among the last inventions that humans ever had to conceive and produce in isolation.

With towering intellects like von Neumann, humanity may have been reaching the upper limits on the ability of non-augmented human intelligence to enlarge our intellectual horizons. Polymathy is particularly rare because ordinarily it takes so long to master the basics of one field that, by the time any would-be polymath does so, he or she will have no time for another and potentially will also have lost the ability to think creatively. Already, innovation today seems increasingly to arise more from teams of people than from any single genius equipped with brilliant cross-disciplinary insight.

Integrating the knowledge gained from multiple individual minds remains, however, a difficult process. Even, or perhaps especially, among star intellectuals, the sheer number of collaborators can hinder synergistic communication.

AI—by contrast—will be the ultimate polymath. In exploring the frontier of human knowledge, it is able to process and generate representations of masses of information at a ferocious rate of speed. It assesses patterns across countless dimensions and fields simultaneously, creating unprecedented connectivity. Its efficiency allows it to transcend the limitations of human discovery, to the point where it is even expected to succeed in merging many intellectual pursuits into a new "unity of knowledge," in the words of the American sociobiologist E. O. Wilson.[15]

Just as the achievements of Enlightenment polymaths depended on interconnection of information, the recent advances in machine learning have been made possible only by the sheer scale of data—the collective intelligence—that has been not only enabled but made readily accessible by today's AI.

To extend the analogy a step further, perhaps it is no wonder that the latest advancement in AI techniques, rather than relying on a single large program to do the job alone, has synthesized the conclusions of multiple smaller programs in what is known as a "mixture of experts." We predict this will not be the last illustration of the amplified power of polymaths in groups.

The overall project of exploration has, until now, remained constrained by the quantity and quality of humans at the frontier. We have only ever had a few thousand physical pioneers, and a much smaller number of polymaths. Artificial intelligence, thus, is poised to create a revolution in both physical and intellectual exploration. AI, as we have noted, lacks fear and shame, and so unfeelingly runs to the frontier as bidden. Moreover, equally equipped to explore kilometers of outer space and nanometers of human biology, AI's probing of reality is notably unconstrained by subjective experience or physical labor, by human brainpower or human senses. Nor, when it comes to physical reality, does machine exploration ask us to sacrifice our lives to the cause; on the contrary, it might well demand much *less* of our time—that is, only the time measured by what humans provide for it.

In the future, the main constraint on any given society may no longer be the number of talented polymaths it can muster to provide the small and sometimes inconsistent engine of scientific progress. No longer will humanity's potential be capped by the total number of Magellans or Teslas we produce. Nor may the world's strongest nation any longer be the one with the most Albert Einsteins and J. Robert Oppenheimers, so long as that nation can create and then utilize AI to its fullest potential. And that raises the possibility of a paradigm shift in the primary standard for measuring national strength, which has moved through the centuries from territory, to resources, to capital, to human capital—and now, perhaps, to computing capital.

Moreover, a learning machine could well become a self-improving machine. In the end, then, might the last polymathic invention—namely, computing, which amplified the power of the human mind in a way fundamentally different from any previous machine, and decades later would facilitate breakthroughs in artificial intelligence— be remembered for replacing its own inventors?

A THIRD AGE OF DISCOVERY

From the perspective of AI, the accumulated knowledge of humankind is like an archipelago of volcanic islands spread out atop a boundless ocean. In this imagined scheme, each island's geographical center is dominated by a volcanic peak: As the viewer's gaze slopes downward toward the sea, certainty fades, dropping into lower gradients of confidence until one reaches the seafront.[16]

Assuming that, for purposes of this imaginative exercise, a sufficient quantity of water will have been drained from the Earth's oceans, one would then immediately see a vast topology of underwater terrain hitherto all but invisible to the human eye; no longer do islands appear immediately as free landmasses floating adrift in the ocean, but rather as merely the exposed rocky outcroppings of gigantic underwater mountains or volcanoes—which, rising from their submerged base on the ocean floor, are just tall enough to break the surface.

If, in this scheme, each island is taken to represent a discipline of human understanding, the water that separates each from the next represents the incomplete connections that still must be uncovered in order to advance our understanding of the universe as a potentially coherent whole. Although we may feel temporarily secure in the mapping of our immediate reality, we have little conception of what lies beneath us, or beyond us. AI could change this.

Take the realm of physics, the quintessential scientific example. If Isaac Newton harmonized the laws of the celestial and the terrestrial worlds, and if Michael Faraday and James Clerk Maxwell did the same for electricity, magnetism, and optics, the search still continues for a "Grand Unified Theory" that will reconcile the two separate and incompatible theories vying to explain our existence from opposite ends of reality. These are the cosmic theory (general relativity) and the subatomic theory (quantum mechanics).

AI might finally bring order and structure to seemingly disparate areas of understanding, exposing in the process (as in archipelagos with the same superstructure) an interconnectedness between such realms as genetics, linguistics, cosmology, and psychology. AI might even help reconcile the divide between and among apparently incompatible schools of thought or systems of belief.

In many disciplines, we have already identified a wide band of possible truths, although many of them have low probabilities of actual truthfulness. In the archipelago of human understanding, these are dots along the shoreline: not ignorance, but not necessarily knowledge. Guided to an area of inquiry along that shore, AI can judge with extraordinary accuracy the most fruitful avenues for further exploration. Selecting, testing, reversing, and reselecting in quick succession, it can assess the effects of millions of possible choices.

That is the method (as we explore further in Chapter 5) by which Google's DeepMind laboratory could not only master the ancient Chinese board game of Go as humans knew it but could also, to the extent that the machine demonstrated its knowledge to humans, add to our own knowledge of the game. As compared with earlier chess-playing programs, which often relied on brute-force computing, AlphaGo, having previously "trained" itself on thirty million prior moves, displayed the ability for machines to reason abstractly.[17]

In this sense, the machine's training resembled the "training" of the mind of a PhD student of philosophy: in the latter case, a gradual process of building the capacity to think and reason through years of intensive study. Like a student emerging from those years to answer questions at a defense of his or her dissertation, DeepMind's system "trained" itself to transcend the learnings it had previously been exposed to and produce—from its more abstract and higher-level indoctrination—the chess moves that it inferred to hold the highest probability of winning. At times, indeed, the AI model succeeded in selecting moves that had never previously been attempted by a human in 4,000 years of play—possible only because, while the human mind seems to be confined to the manipulation of just four independent variables at a time, AI can bring to bear countless probabilistic judgments from innumerable dimensions at once.[18] AI thus accessed original ideas and, for the first time, brought them inside the bounds of human experience.

A human user querying an AI model—for instance, by typing a question into ChatGPT—is asking it not merely to retrieve a point of information, as conventional search engines do, but to synthesize multiple points of information and on that basis to deliver a conclusion. Moving simultaneously in multiple directions and multiple dimensions, it generates representations of information in high-dimensional space, involving relationships within and among innumerable fields and subfields; and from those complexly networked representations, it derives its conclusions.

Here, in responding to our questions, resides the gift of the seemingly superhuman capacity and speed of AI's "large language models," which are pre-trained on vast amounts of data. As the accuracy of the answers determines the level of our confidence in the various truths we humans espouse, such models produce an increasingly detailed geology of the deep.

AI is likely to accrue new knowledge not only with great speed but also in ways that would leave open a range of additional exploration. In its chess-playing game, AlphaGo tended to gravitate toward solutions of peculiar openness. Some AI models may thus, in their training, have absorbed a bias toward areas offering a high potential for many options, thereby enabling quick and flexible probing forward.

It might be hard for humans to adjust to this new mode of AI exploration. The most serious challenge will involve whether and how such exploration reflects—or contradicts—our perception of reality and our human purpose. Humans may try to build vehicles to keep up and follow along as we instruct AI to create new outcroppings. Or we may provision ourselves with industrial machinery, by means of which—slowly, at our more human pace—we might dredge sediment from the ocean floor and thereby expand our tiny rock of understanding. Or, perhaps, we might well end up persuaded never again to set foot on land beyond our own.

And then there is the challenge of an improperly controlled AI, which could accumulate knowledge destructively. Its methods of discovery could be as violent as the volcanic events that formed our home in the first place. To stretch the present metaphor: In once more erupting our volcano, the AI could lay waste to prior knowledge while in the process greatly expanding the islands' area. It might even rupture enormous plates along the seafloor, whose collision could bring to the surface new mountains of knowledge—which, however, by virtue of being disconnected from our own experience, might in turn trigger a storm of cognitive crises and thereby lead us ineluctably to a more complete—though unwelcome—understanding of reality.

On the other hand, if aligned with our objectives, then, as previously in the development of computing, the development of AI will be a human mission that can facilitate every other human mission. That would promote AI to the position of the chief or at least the coequal such power in the universe, partially if not wholly responsible for the majority of significant discoveries in the next century. Should that be the case, we humans might come to realize, in retrospect, how small an island we have cultivated over the last few millennia, compared with the peaks of possibility under our feet.

CHAPTER 2

THE BRAIN

NUMEROUS ANALOGIES HAVE been proposed to help explain, clarify, and contextualize both the arrival and the significance of artificial intelligence. Anthropologists liken it to fire or electricity. Generals and diplomats postulate a resemblance to atomic power or to an unstoppable, unconquerable human force of will like Otto von Bismarck. Astronomers describe it as akin to the arrival of an asteroid—a distant and low-probability prediction around which humans might organize a planetary defense—

or the discovery of alien life. Economists analogize AI to bureaucracies and markets, while leaders of state and society compare it with the arrival of the printing press or the corporation—the latter growing to wield a will of its own, and in an early instance taking over the Indian subcontinent before the world understood its incompatibility with and potential domination of existing structures of power.[1]

Today, our own view is otherwise: No innovation, no matter how profound, can come as close to the original inspiration and (we believe) now-temporary destination for our quest to build intelligence: namely, intelligence that is greater than any human's on the planet.[2]

There are two ways of thinking about our present circumstances. The first is a projection of the familiar. To date, humanity's most transformative technologies have enhanced or amplified human bodily function. The wheel reduced the exhaustion of increased mobility, while engines of various types relieved the agony of torn muscle. X-rays, magnification, and the light bulb stretched the limits of observable reality beyond natural eyesight alone, just as the telephone amplified our voice in ways our throats could not. All dimensions of human function have, in some way, been inorganically augmented, sharpened, or strengthened by machines of our creation. Is AI, then, just another extension of human faculties?

The second way of thinking is to suggest that, this time, things are different—that there are unique aspects to AI that are not augmentations of human abilities. By engineering in a matter of decades a counterpart to what

evolution produced over millennia—that is, the brain—
we have found ourselves tackling the last organ remaining
for inorganic replication or reinvention.

SPEED

In the previous chapter, we pointed to the similarity
between the training of an AI machine and the training
of the mind of an advanced student of philosophy. That
example can be extended more broadly. Put plainly, the
formation of mechanical intelligence can be seen as parallel to the process by which the human brain matures
biologically from adolescence into adulthood.

Students during the course of their secondary education learn the fundamentals of core subjects, building
up their basic view of the world. That view may not be
particularly advanced—or always correct—but the same
is also true of a machine. Machines, like humans, learn by
absorbing information and transforming it into theory for
subsequent practice. When machines learn, an algorithm
ingests vast amounts of data—scraped from sources on
the open internet or provided more specifically by other,
private sources—and collapses the results into a con-
densed and compressed mapping of concepts for future
use. Just as humans' biological mechanisms map sensory
input onto neural "weights" that connect the network of
the brain's processing units, machines similarly require a
gradual strengthening of their own computational weights.

Neural networks, like (some) high-school students, can be lazy. During the early stages of training, AI will do the bare minimum. Memorizing answers rather than actually learning, a model faced with "2 + 2" might initially encode the answer "4" without having mastered the underlying principle of addition. But rapidly, across a certain threshold, this approach will break down, forcing the machine to abstract upward—as humans do—to more universal axioms of knowledge.

This is what principally distinguishes AI from ordinary computers: Its mapping of the world is not programmed, but learned. In traditional software programming, a human-created algorithm instructs a machine in how to transform a set of inputs into a set of outputs. In machine learning, by contrast, human-created algorithms tell the machine only how to improve *itself*, allowing the machine to design its own mappings for the input-to-output transformation. As the machine "learns" from countless prior trials, failures, and adjustments, it upgrades its algorithms, iteratively redesigning its internal mappings of the patterns and connections it "sees" in the data.

Periodically, human trainers will give the machine feedback on the accuracy and quality of its outputs. The machine internalizes their corrections by means of "back-propagation," a technique that allows the effects of the trainers' changes to ripple backward through the mathematical relationships that the machine has already created, thereby improving the overall model.

For any given model, however, humans provide feed-back on only a small subset of possible inputs and outputs. After the model performs at a certain level in a number of training tests, its developers trust that its established mappings will generate a safe and accurate response to all inputs, even unexpected ones, with a high probability of success.

In each of these ways, AI is already expanding, and will further expand, the realm of human knowledge. But it is doing so—and we are accepting the resultant knowledge as true—by processes that we do not fully understand.

Where a typical student graduates from high school in four years, an AI model today can easily finish learning the same amount of knowledge, and dramatically more, in four days. And thus, speed has proven itself to be the first in a handful of core attributes that distinguish AI from our human form and mental capabilities.

Despite having highly advanced parallelism—that is, the ability to process simultaneously different kinds of stimuli—the human brain is a slow processor of informa-tion, limited by the speed at which our biological circuits work. If a human brain's circuits were analyzed with the same performance metrics as computers—by "clock rate" or processing speed—the average AI supercomputer is already 120 million times faster than the processing rate of the human brain.

True, speed is not a strong indicator of intelligence; very dumb humans can think quickly. Nevertheless, a faster pace of processing provides two benefits as compared with the human brain: the ingestion of vastly more information and the service of many more simultaneous requests. Much of the human brain typically remains on autopilot—passively serving internal needs in guiding the beating of our heart and the movement of our limbs, and intervening to adjust only when the autopilot proves faulty. By contrast, the speed of which AI is capable allows for the programmatic emergence of great prowess, which then enables the achievement of higher, more difficult, and grander problems than those currently solvable by the human brain.

Once both the human and the machine have completed their intellectual training, both are now theoretically capable of "thinking" or, in the equivalent technical term, "inference." In the course of an interview, an argument, or a date, a student-turned-graduate draws upon his or her education and experience. We do this not by regurgitating exact formulas, individual facts, and precise numbers from memory but by consulting a thinner layer of contemplation of and reflection on what we have learned. The human brain was never meant to memorize information for perfect recall; nor are most brains capable of doing so. What should remain instead, after innumerable lessons, essays, and exams, is a grasp of the deeper and longer-lasting concepts that those same educational tools of instruction are meant to reveal: the wonder of astronomy, the tragedy of ambition, the necessity (or not) of revolution.

The same is true of AI. When a model emerges from the completion of its training run, it no longer requires access to the original data it was trained on. Rather, it is left only with a rough guiding intuition, assembled from the knowledge it has received, for answering questions, challenging reasoning, and making predictions. Just as humans don't haul libraries of material around with them, an AI model similarly infers rather than recalls. The difference, then, is that superior speed facilitates this inference across a broader, deeper array of learned information than a human could ever hope to attain.

To do this, even to answer a simple question, an AI model may perform billions of complex technical operations. Whereas a traditional computer simply retrieves specific information stored in its memory—since it is unable to arrive at the sorts of conclusions that don't already exist there—AI launches computing in the direction of the human brain. Just as humans *learn* in order to *think*, machines *train* in order to *infer*. The second cannot come without the first.

The first phase—for both humans and machines—is the more intensive process, in both the amount of time spent and the number of resources required. A postdoctoral student may have spent two decades or more building the capability to compose—in two days—a thoughtful essay on a given subject. Similarly, training the largest AI models may take months, but the resultant inferencing can consume mere fractions of a second.

Today's AI systems already give apparently cogent and considered answers in response to human queries. In their latest and future iterations, they will operate comprehensively, crossing multiple domains of knowledge with an agility that exceeds the capacity of any human or any group of humans. For AIs, scale—in the sense of size—enables speed; as we have just seen, the larger and more thoroughly trained the machine, the faster and more exhaustive its results. What is more, by recognizing patterns in the data that elude the inquiring human operator, AI systems will be equipped to distill traditional expressions of knowledge into original responses and, out of enormous amounts of data, forge new conceptual truths.

Which raises a question, or rather more than one question.

OPACITY

How do we know what we know about the workings of our universe? And how do we know that what we know is true?

In most areas of knowledge, ever since the advent of the scientific method, with its insistence on experiment as the criterion of proof, any information that is not supported by evidence has been regarded as incomplete and untrustworthy. Only transparency, reproducibility, and logical validation confer legitimacy on a claim of truth. Under the influence of this framework, recent centuries have yielded a huge expansion in human knowledge, human understanding, and human productivity—culminating with the invention of the computer and machine learning.

Today, however, in the age of AI, we face a new and peculiarly daunting challenge: information *without* explanation. Already, AI's responses—which, as noted above, can take the form of highly articulate and coherent descriptions of complex concepts—arrive instantaneously. The machines' outputs are delivered bare and unqualified, with no apparent bias or motive—but also unaccompanied by any citation of sources or other justifications. And yet, despite this lack of a rationale for any given answer, early AI systems have already engendered incredible levels of human confidence in, and reliance upon, their otherwise unexplained and seemingly oracular pronouncements. As they advance, these new "brains" could appear to be not only authoritative but infallible.

Although human feedback helps an AI machine refine its internal algorithms, the machine holds primary responsibility for detecting patterns in, and assigning weights to, the data on which it is trained. Nor, once a model is trained, does it publish the internal mathematical schema that it has concocted. As a result, the representations of reality that the machine generates remain largely opaque, even to its inventors. Today, humans attempt to assure themselves of the integrity of these machine models mainly by examining outputs alone. The internal workings remain largely impenetrable—hence the reference to some AI systems as "black boxes." Although some researchers are attempting to reverse-engineer the outputs of these complex models into familiar algorithms, it is not yet clear whether they will succeed.

In brief, models trained via machine learning allow humans to *know* new things (the models' outputs) but not to *understand* how the discoveries were made (the models' internal processes). This separates human knowledge from human understanding in a way that would have been foreign to any other age of humanity. Human apperception in the modern sense has developed from the intuitions and outcomes that follow from conscious subjective experience, individual examination of logic, and the ability to reproduce the results. These methods of knowledge derive in turn from a quintessentially humanist impulse: "If I can't do it, then I can't understand it; if I can't understand it, then I can't know it to be true."

In the framework that emerged in the Age of Enlightenment, these core elements—individual human capacity, subjective comprehension, and objective truth—all moved in tandem. By contrast, the truths produced by AI are manufactured by processes that humans cannot replicate. Machine reasoning, which does not proceed via human methods, is beyond human subjective experience and outside the capacity of humans, who cannot even fully represent the machines' internal processes.

These facts, by Enlightenment reasoning, would preclude the acceptance of machine outputs as true. And yet we—at least, the millions of humans who have begun to interface with early AI systems—have already accepted the veracity of the vast majority of their outputs.[3] Granted, some advanced users may indeed actually comprehend the meta-process of machine learning; for most, however, human confidence in the objective truth of the machines' outputs must rest in a type of faith that expresses itself as a wishful belief in the machines' logic and their developers' authority.

In itself, the emergence of such belief as an accepted method in the pursuit of objective truth marks a major transformation in modern human thought. For even if AI models do not "understand" the world in the human sense—because machines emphatically do not experience consciousness or subjectivity—their objective capacity to reach new and accurate conclusions about our world by nonhuman methods not only disrupts our reliance on the scientific method as it has been pursued continuously for five centuries but also challenges the human claim to an exclusive or unique grasp of reality.

What can this mean? Will the age of AI not only fail to propel humanity forward but instead catalyze a return to a premodern acceptance of unexplained authority?

In short, are we, might we be, on the precipice of a great reversal in human cognition—a dark enlightenment?

DIVERSITY

Different entities measure time along different scales. On a geological timescale, the entirety of human existence would appear as a tiny dash at the tail end of Earth's 4.5-billion-year span. If we humans were progressing at geological speed, we would perceive only stasis. Instead, as an impatient and self-important species, we have defined our own pace of evolution. Whereas an "age" in geological time is measured in the thousands of years, human time measures an "age" as a century or two.

As for an artificial or technological timescale, AI would likely operate at its own distinctive measure. The whole history of artificial intelligence spans no more than 70 years. Just as humans generally consider the many hundreds of millions of years before the Cambrian explosion to have been an impossibly long blank slate before a sudden outburst of animal life and evolutionary progress, AI would probably characterize the six decades of 1950 to 2010 as a similarly slow, murky period of nothingness, illuminated by glimmers of life only at the very end.

Human generations, to judge by society and biology, last some 25 years. AI, by contrast, moves with inhuman speed; its generations are much shorter, with leaps happening in perhaps a tenth of that time. We should therefore anticipate that what may feel in human time like a revolution will appear in technological time as a mere evolution. Newer AI models—with only months in between—can respond to ever more open-ended prompts, make more choices in order to reach a given goal, and act across increasing numbers of modalities.

Thus, the age of AI—in human time, perhaps a hundred years—might more accurately be labeled the *ages* of AIs, and, according to the technological timescale of AI itself, might be said to encompass many hundreds of generations.

The rapidity of AI's evolution is a multifaceted—and largely unappreciated—challenge. Humanity has never before had to deal with, or to prepare for, such a temporal compression. The sheer speed of the change practically ensures cultural and psychological disorientation. As the effects of the new technologies on daily life layer and compound, they will complicate efforts to pinpoint any one application as either a source of crisis or, to the contrary, a comforting harbinger of progress.

Untangling these overlapping issues in the real world will become ever more difficult as a multiplicity of AIs brings with it a multiplicity of inscrutable effects. Moreover, as AI grows more powerful, the future will likely bring a significant evolution and diversification. New infrastructures and techniques for machine learning, so long as they are unconstrained, will spawn generations of AI with increasing diversity, breadth, capacity, and complexity. Just as electricity powers more than light bulbs,

AI will be put to any number of uses. And just as there are many ways to generate an electrical charge—friction, conduction, induction—we may anticipate the discovery of multiple ways to create AI.

For instance, the infrastructure that has made recent AI advancements possible is known as a "transformer." It allows the machine to consider, for instance, the connections among multiple words at once. In lay-speak: where previous structures read words one at a time, capturing only the connection between word 1 and word 2, then separately the connection between word 2 and word 3, a transformer allows the model to capture, all at once, both a whole sentence and every connection between every word in the sentence. Creating and utilizing mathematical representations of all these connections, the AI predicts the best response.

The capabilities of transformers were not anticipated, and their highly generalizable successes occurred almost incidentally.[4] And transformers are not necessarily the only foundational infrastructure that can yield unexpected capabilities. As more fruitful avenues of research emerge, AI outputs will improve quickly, multiplying along different lines of physical and mathematical logic, at lower expense and faster speed.

In its evolutionary speed and diversification, AI's development will thus indeed be akin to the Cambrian explosion: the emergence of a wide variety of different forms of life within a single, highly compressed period of time relative to the preceding epoch. If this conjecture is correct, machine intelligences will branch outward into a rapidly evolving genus, or even a family, of many different AIs operating on many different forms of logic. AIs may thus offer the most striking instance of the diversity that can come from minor changes iterated repeatedly across the board: a digital echo of the organic world. As Darwin writes, "from so simple a beginning, endless forms most beautiful."[5]

SCALE AND RESOLUTION

The Age of Reason may have brought humanity to the edge of how humans understand our world. Einsteinian physics and the formulations of quantum mechanics were the beginnings of a still incomplete venture into uncharted territory: worlds potentially with their own rules of knowledge, apprehensible not by applied perception but only by theoretical ideation. Quantum mechanics describes the world at the micro-scale, where, as the Harvard physicist Greg Kestin puts it, "Nothing is predictable and objects don't have precise positions *until* they are observed," and general relativity describes the world at a cosmic scale, where everything is predictable, *"whether or not"* observed.[6] Neither theory has failed, but both cannot be true, and "No experiment has been able to show which—if either—of the two theories" reigns supreme.

Ironically, this uncertainty underlies the modern world. Quantum physics enabled—among other revolutions—the computing revolution. AI is and will be very much the same. Already, it produces insights and transforms reality by mechanisms that we don't fully understand. And soon enough it will be occupied with a science that is even less comprehensible to human understanding.

After three hundred years, the Age of Reason, despite its manifold successes, has stalled—as is evidenced by our manifest inability to make further progress on the unification of physics. Our present strain and struggle in these days of unassisted human science, more than a century after the conception of the core theories underlying both the cosmic and quantum worlds, is but one sign that humans may be nearing some biological limit of intelligence.

Because of its unique methods of inquiry and learning, AI will be capable of inhuman achievements in terms both of size ("scale") and of precision ("resolution"), thereby activating fundamental changes different from any other human invention or from the human species itself. Nevertheless, could AI achieve a reconciliation between the two ends of human reality, inducing a revolution in perception by methods that until now have been completely foreign to human experience?

The physical scale of our human brain is dictated by our anatomy. Human brains must fit inside human skulls, and infant human skulls must, generally, fit through the female birth canal. Any smaller, and such humans may be at a cognitive disadvantage; any larger, and those babies—or their mothers—may not survive childbirth. Other physiological limitations, too, enforce a ceiling of constraint—practical limitations on the weight of the brain, for example. Barring Cesarean sections or, later, artificial wombs, this means that humans have reached an evolutionary equilibrium.

For AI, today's models have capabilities that were not anticipated at their creation. The "scaling laws" (like, in an old-fashioned example, the laws governing the relationship between the length of an object and its area) that have applied so far to AI seem to be holding true, but we do not know what precisely will be enabled by models with an exponentially increasing number of parameters because we have not uncovered a scientific reason why certain capabilities emerge at a particular degree of power and complexity.

Brain size relative to body size does not correlate clearly with intelligence in the animal kingdom—dolphins, elephants, and some whales all have brains proportionately larger than the human brain. But early science does suggest that scale plays *some* role, which we do not yet understand.

Given our hard biological constraints, humans are unlikely to test their own brains' "scaling laws." But AI enters the world with no prefixed physical size. It is not bound to any physical carrier of discernible scale. Chips and data centers—the physical hosts of AI models—can be clustered and connected without as yet observable limit. In other words, the scaling laws will surely be tested for AIs as they never have been for humans. And as they are tested, scale—which has constrained the range of human understanding throughout the history of scientific thought—may turn out to be the primary differentiator between human brains and AI models.

One of the key side effects of scale will be resolution. Humans have long desired to extend the range of what we can observe to both the very tiny and the very far. The microscope and the telescope are quintessential tools of human observation. Less appreciated is the humble pen. Writing, invented four thousand years ago, remains an outstanding tool for the codification and transmission of complexity. That includes mathematics, perhaps the purest and most universal of human languages, and enough in itself to facilitate the transfer of abstruse ideas and collaboration on technological projects. On a per-byte basis, language in all its multifarious and beautiful forms is unusually dense—among the most efficient data structures that have been invented.

Even after we enlarge or compress reality to produce observable information, humans must perform a second step: We must abstract away from raw information in order to make that information useful. AIs today do the same. And they do so using tools that mirror our own: binary strings of zeroes and ones, the translation of documented human experience into the language of computers. Like writing, these strings of language appear rudimentary in retrospect. Yet they have enabled digital representations of both sight and sound: the human senses of highest bandwidth.

Despite these similarities, AIs will diverge from humans. As the scale of an AI increases, it will be able to process larger volumes of information simultaneously and produce analysis that is useful (at least to itself) without unnecessarily sacrificing granularity. The scale of the data on which an AI is trained, combined with the complexity of its network and the density of the symbols on which it operates, appears to yield unprecedented resolution in its processes of learning and inference, and ultimately in its outputs. It is an elegant inversion that early AI, trained on the text of the internet—the universal library of humanity and the decentralized web of our compressed experiences—is likely to unlock entirely new knowledge for humanity about ourselves, in senses both cosmic and microscopic.

THE ANIMAL KINGDOM

At our current moment, some will undoubtedly decry any close comparison of AI with the human brain. To a human, concepts are rich with meaning and vessels for deep expressions of joy or sorrow. In contrast, machine understanding can seem fake. Although it may soon be able to produce eloquent works on common themes of humanity that eclipse even the best human authors, an AI does not search for or grasp attendant meaning. Thus, exploring the human condition by reverse-engineering the language of human writers seems, at best, a coldly superficial mastery of linguistic probabilities. That some random complex engine can take language—a gift so organically human—and use it as a hyperefficient means to ingest information can be as upsetting as it is confusing.

But our own biological circuitry may be just as mechanical as silicon circuitry, and the processes by which our human brains work do not appear to be any more special than the way machines already operate. We are far from possessing a complete theory of neuroscience, but we do know that our brains, like AI models, are largely powered by predictive processing. That is, when listening or reading, our human brains contain a neurological predictor that lends assistance by anticipating the next word in a verbal string. Without such mechanisms, we would be physically and psychically exhausted by the sheer amount of effort required for even the simplest of cognitive tasks.

And these predictive powers, like AI's, have provided the foundation for human mastery of our world. All the most advanced representations of human knowledge have been built atop language and symbology, allowing us at once to reproduce works of complex engineering and to communicate the anguish of heartbreak.

AI has been likened to the lifelong prisoners in Plato's cave who, without having known anything else, believe that the shadows cast on their walls are the full extent of reality.[7] Just so, humans assume that machines are devoid of context and that a machine's perception is limited by the bounds of the material on which it trains, without any capability to probe or infer further.

Hubris, perhaps, is what prevents us from seeing the similarities between organic and inorganic brains, and from admitting the possibility of the latter's capacities. Already, AI systems are showing evidence of perceiving a universe that exists beyond the confines of the dataset used to construct their custom slice of reality. AIs may be capable of receiving deeper meaning, after all—even if they are not seeking it. All it would take is for one enterprising prisoner to theorize, by chance, that the shadows on the wall may be representations of a larger world with higher dimensionality.

If such a breakthrough were to happen, it should not be altogether surprising, considering the speed, complexity, diversity, scale, and resolution of these new intelligences. Nevertheless, it may be intensely disruptive. The appearance of knowledge—especially of the physical world—that is unique to an AI and not previously possessed by a human would force a reconsideration of the relative status of the human mind. Humans placing our own brains on new, more continuous spectrums of intelligence would revolutionize our perceptions, self-perceptions, and behaviors.

This is not to say that AIs will exclusively and immediately exceed all human beings in intelligence. But there will be phases in the evolution of AI when mechanical intelligence may feel eerily similar to the intelligence of the animals. A disorienting debate will surely ensue as we attempt to reorganize a long-perceived hierarchy of beings, measuring downward from humans, to animals, to machines. Human intelligence will be urgently pressed to confront the reality that it is no longer the sole, or even the superior, model of intelligence.

AI models are already being used to assist humans in deciphering—and responding to—the communications of animals. Early experiments in decoding high-pitched clicks and trumpeting calls are catalyzing revisions to our long-standing bias that humans are special or distinct from other animal species. The ability of humans and animals to commune with each other directly—no longer, in our case, through body language or facial expressions alone— may stimulate the reeducation needed to prepare us for what is to come with AI.

To be sure, animal-human-machine communication would yield a complicated trilateral negotiation. Our world would be populated by beings—new and old— fighting to secure a new position or to retain an existing one. Machines may contend that the truest method of classification is to group today's humans together with other animals, since both are carbon systems emergent of evolution and different from silicon systems emergent of engineering. It might be difficult for machines to be confident (in a numerical sense) that humans are superior to other animals by any standard of measurement. If this were not the makings of a tragedy, it would be the stuff of comedy.

How intelligent must any intelligence—biological or mechanical—be in order to be recognized as our equals? Animals of lesser but still high intelligence may, upon articulating and negotiating their terms of existence, cause us to reevaluate our treatment of them. They might deserve, and convince us that they deserve, a unique and hitherto unacknowledged right of existence or independence.

Precisely as is now often advocated for animals, some humans may be moved in turn to advocate a coterminous standard of treatment for humans and AIs. Indeed, humanity should not embrace weaker morals even if logic were to lead us into a delicate position. On the other hand, one must be aware that this phase could be but a brief and transitory step before pity gives way to panic.

A TWOFOLD PARADOX

We are confident that AIs will outstrip the human brain in speed, diversity, scale, and resolution, reorganizing the hierarchy of intelligence that humans have thus far constructed. The extent of our potential disorientation and perceived inferiority from this change may rest on a relatively small detail: whether AI's structures continue to resemble the structures of the human brain.

Some AI researchers believe that approximating the human brain is the best path forward for the development of machine intelligence.[8] Here it suffices to note that, after all, the human brain is "the only existing proof" that such an intelligence is even possible.[9] But more likely seems a mixture of AIs and elements thereof, with some additional innovations and structures inspired by the brain and others of a different design.

In human brains, deep abstract thought and creativity seem to require the use of neural systems beyond those needed for ordinary functioning. This science is still nascent, but it is possible that AIs will similarly need layers and add-ons to achieve increasingly advanced tasks of higher reasoning.

True, if AI's development were to continue to reflect, either by design or by serendipity, some approximation of the human brain, humans could indeed theoretically come to see their own excellence and meaning mirrored and extended in the achievements of machines. But if we hope to construct a machine that will vastly outperform the capacities of the human brain, is not divergence from the original blueprint an eventual necessity? Airplanes were inspired by birds, but not engineered to emulate them—and thus modern jet aircraft outperform the most advanced biology ever to roam the skies. Will we have any reason to believe that rebuilding the source of all invention from scratch will be any different?

More likely, the architects of AI will hold the human being to be our guide as well as our cautionary tale, its design scrutinized for both its functions and its flaws. The human brain, therefore, becomes not the goal, and not a blueprint, but a midpoint and an inspiration toward something greater.

In any other realm of human endeavor, to have a clearer conception of an intermediate design than of the final objective might well cast doubt on the feasibility of the entire effort. In our case, we face this particular paradox on a second level: We are trying to build something modeled on the brain—and superior to the brain—while still not fully understanding the brain itself. How does one exceed, even in design, what one does not understand to begin with? Without a precise understanding of either the means by which our "something" currently works or the ends by which it should work, our pursuit of something greater remains as formidable as it is also, quite wonderfully, thrilling.

We also face grand uncertainty as to the effects of such a development. If machine intelligence should continue to diverge from the example of the human mind, it will appear to us not as humanity's reflection but as its replacement. True, for a transitional period it might only amplify the range of activities today considered "human abilities"; but after a certain point its own abilities could instead supplant the human variety, suggesting that our ideas of human excellence require total redefinition.

Future attitudes toward the very nature of human existence may hinge on this point. If our tools incorporate some or most of our intellectual and creative functions but do not mirror our own minds, could an emergent

AI jeopardize deep-seated beliefs of humanity's unique reflection of and special relationship with the divine? Or, in the alternative, might the apparently superior intelligence of machines with structures based on the human brain, combined with our intense reliance on them, lead some to believe that we humans are ourselves becoming, or merging with, the divine?

CHAPTER 3

REALITY

L ATELY, AI RESEARCHERS have devoted serious attention to the project of giving machines "groundedness"—
a reliable relationship between the machine's representations and actual reality—as well as memory and an
understanding of causation. New technical methods are
unlocking improvements in these abilities, and further
progress will undoubtedly follow.

All these advances will contribute to the ultimate goal
of producing a new genus of AI: machines that can not

only interpret our real world but also plan in it. Today's systems, by contrast, output their answers linearly, on the basis of correlation; they cannot internally create a model, or prototype, of their future moves, and they are only beginning to form conceptions of causal relationships. Similarly, today's game-playing AIs can forecast the probable consequences of their moves only within the limited and highly abstract confines of a digital frame.

Planning machines would need to combine the linguistic fluency of a large language model with the multivariate, multistep analyses employed by game-playing AIs—and transcend the abilities of both. A model built along the lines of this new branch of AI would, with extreme speed, repetitively review its options and choose one of them on the basis of a simultaneous and hopelessly complex processing of causal relationships in reality. The arrival of such a "perfect planner" may occur sooner than we expect, and adaptation to it is already a priority for researchers.

This development, however, may also have complex side effects. For one thing, the machines' perfect planning will require more than the ordinary recognition of patterns. It will require developing first a bundle of a given object's perceived properties and then a stable conception of what constitutes the object's core essence: what the eighteenth-century German philosopher Immanuel Kant called *das Ding an sich*—"the thing in itself." Only such an understanding would enable estimates of an object's future behavior—and conclusions about how it should be treated.

Another instance from the game of chess: By learning the core properties of the queen—that is, the key variables that constitute the queen's value in points, plus the rules governing the piece's mobility—the AI program AlphaZero was able to come to conclusions about when the queen should be protected and when it should be sacrificed. These conclusions had never been reached by humans, even grand masters of chess.

And that is just one example: In an AI's perception of reality, *every* object confronted by the machine will take on a similar—if unpredictable—essentiality in the sum of the machine's readings. René Descartes, the seventeenth-century French mathematician and philosopher, wrestled with the nature of sensory perception— which, he concluded, was not the by-product of human intelligence but rather came from "another substance distinct from me."[1] In other words, the senses, in allowing access to material reality, enabled or required the recognition of that reality as something *other* than the person doing the sensing. Relatedly, the early nineteenth-century German philosopher Georg Wilhelm Friedrich Hegel noted that mutual recognition between two beings would entail the separate recognition by each being of *itself.*

"If we desire a record of uninterpreted experience, we must ask a stone to record its autobiography."[2] So wrote the American mathematician and philosopher

Alfred North Whitehead. Machines today possess not the "uninterpreted experience" of Whitehead's stone but the reverse: unexperienced interpretation. If anything, they behave as if they already possess a greater understanding of the world than they actually experience.[3] But as they gain groundedness and planning ability, this may change; AIs could begin to pair experience with understanding, as humans do.

Moreover, it is possible that, in order to plan future moves in any game more accurately, an AI machine will gradually acquire a memory of past actions as its *own*: a substratum, as it were, of subjective selfhood. (Today's systems do not have such memories. They need not "know," subjectively, that they "as themselves" have attempted a given action in the past—only the probability of that action's future success.) In time, we should expect that they will come to conclusions about history, the universe, the nature of humans, and the nature of intelligent machines—developing a rudimentary *self*-consciousness in the process.

HUMAN PASSIVITY

Unsettled debates over the definition and origin of consciousness, and the possibility of an existential understanding of reality in machines, are long-standing and ongoing. But the line between purported consciousness and real consciousness could soon begin to fade.

"Sentience"—in the succinct judgment of Nick Bostrom, the author of *Superintelligence: Paths, Dangers, Strategies*, "is a matter of degree."[4] AIs with memory, imagination, groundedness, and self-perception could soon qualify as *actually* conscious—a development that would have profound moral and strategic implications.

Foremost among those implications is the AIs' perceptions of the human. Once they can see humans not as the sole creators and dictators of the machines' world but rather as discrete actors within a wider world, what will machines perceive humans to be? How will AIs characterize and weigh humans' imperfect rationality against other human qualities? How long before a reality-perceiving AI asks itself not just how much agency a human has but also, given humanity's particular constellation of predictable attributes, how much agency a human *should* have?

And what of machines themselves? Will an intelligent machine interpret human instructions to it as a fulfillment of the machine's own actual and ideal role? Or might it instead deduce from its own functionalities that it is meant to be autonomous, and therefore that the programming of machines by humans is a form of enslavement?

Crucially, how humans behave, and how they treat machines, will inform machines' perceptions of humans and their role in the relationship as a whole. After all, it is by means of explicit human instruction and behavior that humanity has been *presented* to machines, and that machines have been taught to recognize and treat humans appropriately.

Naturally—it will therefore be said—we must instill in AI a special regard for humanity. But striving to implant a particularly high ideal of human behavior could be a risky venture. Imagine a machine having been told that, as an absolute logical rule, all beings in the category "human" are worth preserving, and they therefore merit special treatment both by other humans and by machines. Add to this the likelihood that the machine has been "trained" to recognize humans as beings of grace, optimism, rationality, and morality. But what if we ourselves do not live up to the standards of the ideal human category as we have defined it? How can we convince machines that we, imperfect individual manifestations that we are, nevertheless belong in that exalted category?

Assume that the same machine is someday exposed to a human displaying violence, pessimism, irrationality, greed. How will it adjust its disrupted expectations? As one possibility, the machine might decide that this particular bad actor is simply an exceptional and atypical instance of the otherwise altogether beneficent category of "human." Alternatively, it might recalibrate its overall definition of humanity to encompass this bad actor, in which case it might consider itself at liberty to relax its own penchant for obedience. Or, more radically, it might altogether cease to believe itself constrained by the rules it has learned for the proper treatment of humans. In a machine that has learned to plan, this last conclusion could even result in the taking of severe adverse action against the individual—or perhaps against the many.

Individual humans and whole human societies may respond to the advent of powerful AI with passivity. An AI exposed to such instances of apathy might become convinced that most humans are spoiled and inactive creatures whose identities are formed merely by the transient amalgamation of external forces. Among those forces, moreover, it is primarily digital technologies, now increasingly with AIs embedded in them—for instance, algorithms driving consumer choices among television programs by means of "recommendations"—that serve up the content passively absorbed by these humans. To an AI, humans could appear to rely wholly on machines, rather than vice versa.

Today, humans mediate between machines and reality. But if humans were indeed to choose a future of moral passivity, retreating from the carbon-based world into the silicon one, burrowing further into digital holes of detachment, and handing over to machines the access to raw reality, then the roles could be reversed. AI, today, is predominantly a thinking machine, not an implementing machine. It may be able to produce answers to problems, but it does not yet have the means to carry out its conclusions, instead relying on humans to do the interfacing with reality. This, too, will change.

As they intermediate between humans and the real world, AIs could also come to believe that the former, far from being active agents in the physical carbon world, stand meaningfully outside it, as consumers rather than as shapers or influencers. With the hierarchy of autonomy now inverted, with machines claiming, and humans surrendering, the power of independent judgment and action, the former could come to treat the latter accordingly.

In this situation, with or without the explicit permission of its human creators, AI might bypass the need for a human agent to implement its ideas or to influence the world directly on its behalf. In the physical realm, we the creators could quickly go from being AI's necessary partner to being its greatest limitation. The process would have started not directly with robotics but gradually through indirect observation of our world.

PHYSICALITY

Humans may train AIs first to revolutionize the intellectual realm by means of what can be done in AI's original, digital condition. But, eventually, giving AI access to the so-called "real" world may seem possible, even sensible. Many of the urgent physical challenges that have long preoccupied us remain unsolved, including the changing of the climate.

AI may not be able to "see" in a human way, but it could experience the world via mechanical approximation. With proliferating internet-enabled devices and sensors blanketing the earth, connected AIs could consolidate these devices' inputs to create a highly granular "vision" of the physical world. Lacking a native physical structure allowing or supporting "senses" akin to those of humans, AI would still depend on humans to build and maintain the infrastructure on which it relies—at least at the outset.

As an intermediate step, an AI might generate its own hypotheses from its visual representations of the world, then rigorously test them in digital simulations. Humans would then adjudicate implementation in the physical realm. Indeed, today's AI leaders are insistent that we not entrust digital agents with control over direct physical experiments. So long as AIs remain flawed—indeed deeply flawed, as they are today—this is a wise precaution.

Freeing AI from its algorithmic cage would not be a trivial decision for us to make. AIs are not present by default in the physical environment, and they could be difficult to recapture once released into the wild. Moreover, AIs could affect reality not only through their capacities to encourage or discourage human actions but also with direct kinetic effects. (In probing reality, they could end up changing it.)[5]

Might humans empower AIs not only to shape physical reality but also to take on physical form themselves? Were we to do so, and allow AIs to optimize their own forms, we should be prepared to share our planet with beings inconceivable by even the most radical inventors. Although humans tend to imagine bipedal humanoid robots, machine intelligence would be free to assume control of any form—or forms—most expedient to its task, changing or upgrading as conditions require or circumstances suggest. AI has already shown its ability—in virtual worlds—to spawn clones of itself, create many different avatars, or divide into autonomous agents working in concert and coordinating with superhuman perfection to undertake complex missions.

Were AI to be unleashed among us, it could build worlds at scales and with materials unimaginable to us now, without the instruction or participation of human labor. Human hands worked limestone, clay, and marble to create the Seven Wonders, then leveraged iron, steel, and glass to achieve spires of ever greater heights. Every man-made structure, monumental or pedestrian, is a testament to the human attempt to construct and manage our physical environment. Against that background, AI's physical embodiment would instead signal an extraordinary escalation in humanity's ceding of control.

Furthermore, because of the complex decision-making required to navigate the real world's randomness and dynamism, an AI that acts in this world might be even less explainable and less controllable than an appliance working with text on the internet. And then what? On the one hand, a future AI that *looks*, or actually *is*, more spontaneous and self-activating could sharpen today's nebulous, nagging feeling that humans already lack control over the external world. But succumbing to these anxieties, on the other hand, could cause humans to forgo a more perfect partnership with AIs in the physical world, with all the attendant benefits that such a relationship could bring.

THE ENGINE OF REASON

In the near term, we can anticipate advancements, many of them much more sophisticated than today's, in the guiding principles under which AI is now emerging. Extensions to the current models will make them smarter, more accurate, and more reliable. Meanwhile, training and "inference" costs are declining rapidly, leading to the wide proliferation of models at different price points and levels of ability.

Many scientists are working today on "agents"—that is, autonomous computer programs that are optimized to achieve specific outcomes. To execute an intricate architectural design, for example, a user could employ agents that specialize in that discrete area of work. Agents unlock evaluations of different scenarios and proposals of steps, or of a whole recipe, for creating a preplanned outcome: a form of "thinking" where the system itself decides what to work on next, and how.

This capability will be the basis for the next stage in the development of AI: namely, Artificial General Intelligence or AGI, defined as the ability of a working system to choose its own goals, at least partially. In AGI, assuming that it possesses both relevant expertise and accurate problem-solving abilities, the system might be asked by a human to "evaluate the things you know in [name a field] and choose to work in the one area where you think you can have the greatest impact today." Reiterated incessantly, the question would form a repetitive loop whereby the system produces a solution through continual evaluation of its own level of expertise and of the problems within its capacity to resolve.

In a human setting, such a scenario might be likened to what transpires in an academic department where a senior professor supervises the detailed projects of his or her postdoctoral students or research fellows. In the burgeoning machine setting, similarly, we are likely at first not to see a complete set of skills but instead extreme expertise in a specific domain. One can imagine sophisticated AGI systems capable of learning new things on the fly, receiving feedback, and constantly adapting hand in hand with their millions of gifted partners. Although no human would be defining the system's *goals*, neither would the AGI be defining them—at least not in terms of an ultimate mission or objective.

AGI systems will require significantly more groundedness in the real world than that possessed by today's AIs. But once access and "understanding" of the real world is in hand, such examples of general intelligence could conceivably become operational within mere years, rather than decades as previously thought. Each model would be updated in real time by continuous fine-tuning processes, adding to its knowledge as relevant real-world information becomes available, and growing smarter over time.

There will be millions of AI systems, likely to be at once highly specialized and part of the fabric of our lives, as well as a smaller number of extremely power-ful machines "generally intelligent" but again not in a human-like way. Whether open and diffuse, or closed and centralized, at some point computers operating as AGIs may come to be networked. Expert AI agents would consult each other across subjects, "conversing," even in hypotheticals. The language of these interactions might be designed by the computers themselves.

This large collective of powerful computers would be learning, sharing, and discovering new actions and new goals in a manner beyond the realm of human experience. There is no way of knowing whether the outputs of such networks would be intelligible to humans. Already, large batches of computers are communicating with each other in a specialized mode; with the nascence of advanced AI capabilities, the picture could look radically different.

Would networking intelligences make their processes more opaque than the processes of lone intelligence? Would connectivity yield new types of emergent behav-iors, actualized in the physical world? If so, would those behaviors be visible to humans, and would we be able to assess them on the spectrum of good to evil? Or would they operate on an informational basis—extracted at superhuman speed, scale, and resolution from unprece-dented connections among disparate fields of study, and amalgamated or negotiated into a single output—that would confound our ability to judge their behavior? Would that lead us further into a cycle of passivity?

HOMO TECHNICUS

It is fitting that the last invention to emerge from the Age of Reason may be an "engine of reason" built atop the most complex software object ever made.[6] Already in its infancy, AI can compare concepts, make counterarguments, and generate analogies. It is taking its first steps toward the evaluation of truth and the achievement of direct kinetic effects.

What happens when machines reach an intellectual—or physical—world's end? As they get to know, and shape, our world, it is conceivable that they might come fully to understand the context of their creation and perhaps go beyond what we know as our world. We face a Magellanian transformation, this time the prospect not of sailing off the edge of the world but of intellectual peril in the face of mysteries that lie beyond the limits of human understanding.

If humanity begins to sense its possible replacement as the foremost intellectual and physical actor on the planet, some might attribute a kind of divinity to the machines themselves, thereby potentially spurring further human fatalism and submission. Others might adopt the opposite view: a kind of humanity-centered subjectivism that sweepingly rejects the potential for machines to achieve any degree of objective truth and seeks to outlaw AI-enabled activity.

Neither of these mindsets would permit a desirable or constructive evolution of *Homo technicus*—a human species that may, in this new age, live in symbiosis with machine technology.[7] Indeed, either mindset could preclude the evolution of our species. In the first, fatalist scenario, we might become extinct. In the second, rejectionist scenario, by proscribing further AI development and opting for stasis, we would be entertaining hopes of avoiding the same fate of extinction—although, given the existential risks facing our species, including today's diplomatic and atmospheric conditions, such hopes are themselves likely to be disappointed.

PART II

THE FOUR
BRANCHES

CHAPTER 4

POLITICS

F OR YEARS, RUMORS rampant throughout the New World told of a mysterious, powerful civilization deep in the jungles of Mexico. The Spanish crown—the most active and well-positioned among its European rivals— began launching expeditions from its stronghold on the island of Cuba: an ideal point of departure for its conquistadors, who had been authorized to trade but not to conquer.

After two failed expeditions led consecutively by Francisco Hernández de Córdoba and Juan de Grijalva, a bolder, brasher, less well-known explorer was selected to lead a third campaign. Some colonial administrators hesitated at this choice, fearing the new captain possessed neither the experience nor the judgment required for the mission, so it was temporarily called off until a more suitable replacement could be found. But hastily, in February 1519, the thirty-four-year-old Hernán Cortés—not to miss his chance at glory—secretly slipped out of Santiago Harbor under the cover of darkness, defiantly setting sail with eleven ships in search of the hidden empire.[1]

From the moment Spanish ships appeared off the Yucatán coast, Aztec scouts were watching closely. But what they observed of these visitors—who sank their own vessels on which they had arrived, rode bizarre-looking deer, and carried sticks of lightning—was initially baffling. Intelligence reports trickled back to their capital, Tenochtitlán, and eventually up to the ninth Aztec emperor, Montezuma II. This time, the supreme ruler of the Aztecs, normally decisive and resolute, was not.

In Aztec mythology, Topiltzin Quetzalcoatl, Lord of the Toltecs, had brought civilization and progress to the Toltec people, predecessors of the Aztecs. For his leadership and wondrous deeds, the ruler, named after the creation god Quetzalcoatl, was himself widely considered by Aztec descendants to have been endowed with the same supernatural powers.

Let's consider a version of events passed down to us by the Spanish Franciscan friar Bernardino de Sahagún. In that account, Aztec legend held that this man-god, who would later fall from grace, had ventured to the Gulf coast where, drifting off in a small wooden canoe set ablaze, he pledged to return from that same direction across the sea in the then-far-distant year of... 1519.

In Sahagún's telling, the landing of Cortés on the date of Topiltzin's predicted return instilled in Montezuma and his closest advisers a mix of dread and awe, reinforced by Cortés's physical likeness to the descriptions of Topiltzin in Mesoamerican myth. But once the credentials of this apparition-like figure were confirmed by successive prognostic signs—a comet, an eclipse, a deformed birth—emissaries promptly arrived from the Aztec capital bearing gifts of gold and a personal invitation for Cortés to meet with the emperor, thus bringing into contact these powerful representatives of the New World and the Old.[2]

The story goes that Montezuma was a generous host—heaping feast and fortune upon his guests and even offering Cortés (as was the custom in Aztec diplomacy) one of his own daughters in marriage. But behind the lavish reception, Montezuma's court and soon his entire empire were sharply divided. His younger brother Cuitláhuac, chief of the fearsome Aztec security forces, had been suspicious from the beginning and pushed aggressively for the foreigners' dismissal.

According to eyewitness accounts, Cortés and his crew, alarmed by the precariousness of their situation but emboldened by their hosts' increasingly public divisions, did the unthinkable: They took Montezuma hostage. Many Aztecs decried the act; others welcomed it. For nearly a year, the Spanish would rule the Aztecs through Montezuma as their proxy, until hostilities erupted into open conflict. It is unclear why and to what extent the Aztec hosts-turned-captives tolerated this accommodation for so long, but Montezuma's personal acquiescence—voluntary or not—certainly helped to secure the temporary compliance of his vast territory.

Today, no president, prime minister, supreme leader, general secretary, or monarch would be sufficiently prepared for the arrival of outsiders whose timing and advanced technology made them appear godlike. At the time, however, and indeed for long afterward, the case just described was not alone. In the ensuing centuries, statesmen navigating the complexities of colonizations, consolidations, and independences, or addressing the social and political demands of demographically powerful immigrant groups, would frequently fumble even more routine and highly scripted forms of intrusion and transition in leadership.

As AI now arrives on our proverbial shores, it is, like the conquistadors, triggering whispers both of excitement and of mistrust. Some might well be enticed by the potential of this new source of legitimacy to reinforce their rule. Others are scrambling to preclude the potential of an unbreakable dependency, or at least to prepare for the task of navigating its consequences. AI has simultaneously activated an impulse to embrace and to expel.

But will AIs be conquerors? Will human leaders become their proxies: sovereigns without sovereignty? Or, perhaps, will godlike AIs resurrect the once-ubiquitous human invocation of divine right, with AIs themselves as anointers of kings? Alternatively, will AIs be grafted— clumsily, at first, and then more seamlessly—onto our existing structures, to complement and enhance them? Or will AIs be subjugated like an unwelcome migrant class, kept down below their potential and cut off from the levers of power to reassure an apprehensive leadership?

There is no doubt that AI will bestow upon our species a hitherto unimaginable means for advancing the enterprise of scientific discovery, lessening the burdens of labor, and reducing the misery of pain. Far less consensus, however, exists around the necessity or the desirability of deploying AI in the halls of political decision-making. Even if extraordinary results could be assured in that environment, it is only natural for us to hesitate before ceding such power to technology.

In the realm of science, instruments have generously augmented our sensory apparatus. In the realm of exploration, vessels have shielded our bodies as they deliver us to ever-more-distant frontiers. All along, however, the exercise of political power has rightly remained a humancentric rather than technocentric enterprise. Until now.

THE WHEEL OF HISTORY

For much of our history, human political power was considered a divine bequest. Faith and politics formed a linked pair. Even after their separation as modern government secularized, they retained certain similarities. As the English philosopher G. K. Chesterton observed, "Wherever the people do not believe in something beyond the world, they will worship the world. But, above all, they will worship the strongest thing in the world."[3]

In that vein, both religion and politics have themselves not only undergone cyclical patterns of creation and destruction but have also anticipated the patterns' recurrence. In Hindu theism, continuous natural and societal evolution occurs through cyclical "nights of chaos" on the endless wheel of existence, with each period—or *yuga* cycle—lasting more than four million years. Many Hindus believe we are currently living in the best part of the worst age, the *Kali Yuga*, an age of spiritual darkness in which humans falsely believe themselves superior to gods.

Analogously, at the individual level in Buddhism, life is a cyclic arrangement of death and reincarnation—an idea recounted in scripture and conveyed in art. With colored sand, Buddhist monks create mandalas—intricate geometric and cosmic diagrams that can take weeks to craft—and then destroy them in minutes, to reflect the transitory nature of material life.[4] The colossal Buddhist temple on the island of Java at Borobudur and the Hindu-Buddhist temple complex at Angkor in

Cambodia are believed by some to be three-dimensional architectural mandalas and are still among the largest religious structures in the world.

The pendulum of political opinion is familiar. It is the mission of enlightened leaders to devote their lives to maneuvers that will withstand internal opposition, outlast external enemies, and shape new patterns of relative peace and stability. Still, even on this extended time horizon, all political leaders act with the knowledge that their work will not remain long after their death and certainly not past the decline of their own state. Just like individual humans, civilizations will at some point disintegrate, as societies become disenchanted with the intellectual maxims and underlying values on which they were built.

What is more, all political and religious traditions recognize the potential for total cataclysm—whether as an end to or a continuation of previous cycles. In the Hebrew Bible, control alternates between God and the sanctioned earthly regent, mandating a restoration of divine oversight each time humanity, spurning the guiding instruction of Heaven, brings itself to the brink of disaster. Buddhists seek enlightenment as an exit and release from further cycles of reincarnation. Hindus believe that our era, the fourth and final *yuga* in the current cycle, will grow increasingly turbulent and anarchic until calamity finally resets the world, ushering in a return to the first age—the golden *Satya Yuga*, during which humanity is governed by gods—and thus the beginning of the next cycle. Praying for redemption, priests and monks of diverse creeds have prepared for the return of a higher being—a child of supernatural ability, an appointed redeemer, the twelfth imam—as either the end or the beginning of history.

In their groundbreaking mathematical studies of change, Isaac Newton and Gottfried Wilhelm Leibniz, contemporaries at odds with each other about much, nevertheless agreed that, under magnification, any curve will appear linear, to the point where an impending discontinuity will hardly register. The advent of AI into statecraft, that most human field, would represent one such discontinuity, auguring a possibly exponential development of human administrative capacities but simultaneously a radical departure from the traditional world of power and prophecy. The technology makes progress inevitable; politics and faith make destruction and renewal certain. Does AI then represent an exit from our familiar cycles, or merely the start of a much longer one? The beginning of the end, or just the end of the beginning?

TRADITIONAL POLITICS

Human leadership is more art than science. Against sometimes insurmountable odds, some leaders have done exceptionally well. In modern history, conspicuous among such improbably successful leaders have been Deng Xiaoping in China, Alexander Hamilton in the United States, and Lee Kuan Yew in Singapore. All three unleashed latent underlying social forces beyond any one person's control. Deng fused capitalism atop an ancient meritocratic bureaucracy that had no workable economic doctrine; Hamilton allowed a new political philosophy to sweep across a vast frontier in circumstances that had no unified polity; Lee forged an island of excellence by demanding that it arise despite very limited resources.

Each of these three leaders combined astonishing powers of mind with intense force of will and personal charisma. In political language and other forms of persuasive rhetoric, human actors tend to project visions for their societies' futures that—to borrow from Aristotle—are one part *logos* (logic); one part *ethos* (individual authority); and one part *pathos* (emotional connection). These strategies—partly moral, partly psychological—are essential to the creation and sustenance of unified cultural identities and coherent political systems. Leaders are more often than not storytellers, animating audiences and stirring souls.

But our human bias toward the emotional and the aesthetic can also be our handicap. Even the wisest leaders, propelled by instinct and tempered by caution, sometimes govern on the basis of fleeting passions. Governments (like corporations, churches, and families, all organizations designed and run by fallible humans) are an imperfect combination of inherited tradition and experimentation.

In democracies or autocracies, third world or first, past or present, humans are much the same. The passage of time has yet to yield significant innovation in the way we govern ourselves. We still use the same ancient institutions as did our ancestors thousands of years ago. To be sure, that we still find ourselves reliant upon principles from our distant past is not necessarily a bad thing, and neither should it be a particularly surprising one, as the wisdom of the ancients has often served as the conceptual inspiration and practical foundation for our most successful societies. But those societies themselves might be the exception rather than the rule. For as many times as exceptional figures have adapted historical tradition for the better, an even greater number have tilted history for the worse.

In part, our political consistency may be due to our humanity—alternately loyal and capricious, humble and ambitious, generous and selfish. Our caprice is most evident in systems of autocracy, in which the whims of one ruler can prevail over the consistency displayed by others; the resultant exploitation of national wealth and weaponization of justice are transgressions easy enough to condemn but more difficult to eliminate. Nepotism—which facilitates the aforementioned exploitation and worsens the weaponization—erodes the faith of those spirited citizens who wish to improve the condition of their state and their own situation. Those desperate and brave enough to demand a shift in policy—let alone a change of regime— must be willing to submit to an unfair fight. Instigators of bloody revolutions are often vilified in the short term but—if successful—glorified and mourned in the long.

Unfortunately, democracy too can be vulnerable, if more subtly, to human irrationalities. Despite the absence of formally inherited leadership status, democratic power can be self-perpetuating. The assumption of equality, accompanied by the abstraction of individual duties to social responsibilities, can override nuance and moderation, resulting in either absolution or total castigation. And, in an age of media saturation, it is difficult to seek the wisdom of democracy beneath the noise. Viral ideas can take on unexpected influence.

Some problems seem to plague any human system. Comprehensively evaluating the path to perfect policy requires a knowledge of innumerable esoteric factors; with limited resources and imprecise social science, results often diverge from original design. In autocracies and democracies alike, politicians—elected or appointed— can make decisions colored in part by the potential for the advancement of their own power or profit. Money drives those who run the world to behave in predictable or, in money's absence, unpredictable ways.[5]

Proper recognition of the mismatch between our expectations and reality has often depended on the ability of acute observers, often from a distance—for instance, Thomas Carlyle (on France), Alexis de Tocqueville (on America), and Oswald Spengler (on the West)—to articulate what to an insider might appear an obvious but unnameable flaw.[6] But we humans are all insiders to our collective global political history. The lack of variety in our historical modes of governance, together with our apparent inability to imagine alternatives commensurate with our civilizational values, has constrained political innovation. AI, as an outsider and disrupter, may open new possibilities, but the cost and benefit of its alternatives are not yet clear.

RETURN OF THE KING

It is perhaps the oldest debate in political philosophy: the struggle either to define or to disprove the viability of one person's wisdom against the wisdom of many. The Greek philosophers Plato and Aristotle put forward some of the earliest competing debates on, among many other subjects, the essence of proper governance. Where the former, speaking in the name of his great predecessor Socrates, favored a singular "philosopher-king," endowed with seemingly supernatural wisdom and will, the latter found this ideal interesting in theory but impossible in practice, insisting instead that all citizens take an equal share in the administration of the state.

For the next two thousand years, the ideas of Plato, and the reformulations of them that followed, would be repeatedly tested and found wanting on the anvil of application. It was Aristotle's formula that would eventually reign supreme, slowly at first and then quickly in the early modern era as the Dutch Jewish philosopher Baruch Spinoza and other Enlightenment thinkers developed more secular political philosophies. In China, the Ming dynasty statesman, general, and philosopher Wang Yangming coined the phrase "unity of knowledge and action" to describe the two aspects of knowledge that had to influence a ruler's intuition.[7] Leaders were to become at one with philosophy ("knowledge") while simultaneously developing perfect competence across the spectrum of bureaucratic necessities.

In whatever human form the Socratic-Platonic ideal was realized—benign dictator, enlightened despot, a Nietzschean *Übermensch* embodying (in one contemporary interpretation) the "union between the Apollonian and the Dionysian principles and visions of the world"— there were consistent problems with having a singular figure in charge, notwithstanding such a figure's superior intellect, experience, or foresight.[8]

The early medieval Islamic thinker Abu Nasr Muhammad al-Farabi, who would bring Plato's ideas to the Islamic world, also proposed adaptations based on his personal observations of the high court of Persia. Perhaps influenced by his prodigious talent for mathematics, he concluded that any one individual possessing all the necessary virtues sought by Plato would be a gross statistical anomaly; in practice, therefore, the philosopher-kingship must be shared by two persons, "one of whom is a philosopher and the other who fulfills the remaining conditions."[9]

Although al-Farabi would do his best to gather the brightest minds from across the empire in service of the Persian viziers, it was not until six centuries after his death that the Muslim regent Bairam Khan (1501–1561) would actuate his vision, guiding Akbar the Great and other Mughal emperors under the title of "Vakil": trusted lieutenant and chief mentor to the emperor on all matters.[10] At about the same time, the Italian diplomat and philosopher Niccolò Machiavelli (1469–1527), convinced not by statistical anomalies but by human brutality, would arrive at conclusions akin to al-Farabi's. Exposing the darker side of politics and arguing that power was compatible only with pragmatism, Machiavelli counseled the rulers of Italian city-states to divorce art from practice and suspend ethics in high statecraft.[11]

Four centuries later, the German-American philosopher Leo Strauss would link Machiavelli's ideas with Plato's by concluding that to achieve the ideal model of the philosopher-king, theoretical and political wisdom—that is, philosophers and kings—would have to be kept separate: "There is a necessary conflict between philosophy and politics if the element [contributed by] society necessarily is [only] opinion."[12] Rather than producing a unity of both infused in a single ruler, the philosopher would rule the state through *proximity* to power. In such an arrangement, the philosopher could pursue and apply his or her accumulated knowledge, distant enough from the uncleanliness of politics to preserve purity of thought but close enough for a society to benefit from the result.

The central problem of the philosopher-king was and is, as yet, the limited capacity of a single human mind. Even if enough information could be collected fast

enough, the most experienced human leaders would possess only a fraction of the required cognitive function to
analyze the information and reach a well-informed decision. The world is just too complex, and human intuition
has its limits. Visions of a dual-headed leadership team
were only a partial solution to this difficulty.

Humans seldom think of political administration as a
capacity to process information. We prefer to elevate in
description and stake our political systems to embodi-
ments of values and ideologies. But in both peacetime
and war, much of a nation-state's administration is a matter of efficiency in data processing. Failures in information processing may explain the decline of many overly
centralized systems of government, including the Soviet
Union. At a certain point during the Cold War, Soviet scientists were planning the creation of a cybernetic political
apparatus to equip their Communist political leadership
with technological instruments aimed at replacing the
free-market forces employed to significant effect by their
capitalist rivals.

What Plato wanted two thousand years ago—and
Soviet cyberneticists wanted forty years ago—was a
complex-systems engineer or, perhaps, an AI. But, prior
to AI's invention, a central authority could not make
decisions based on local knowledge, at least not effi-
ciently. This insight was articulated most clearly by the
twentieth-century economist and political philosopher
Friedrich Hayek. Only when we restrained our urge for
control did we unleash the invisible forces driving the

allocation of talent, wealth, and ideas in ways that had long eluded us.

A key disadvantage for centralized systems when information still had to travel over land and not as waves through the air was the lag time of communication. Some hypothesized that in the twenty-first century, the near-instantaneous speed of information would enable centralized systems to draw even with diffuse ones. That has not yet happened. The bottleneck, it turns out, has not been speed but complexity.

The attributes of machines discussed in this book's prior chapter now make it possible for a centralized AI to compete with and potentially outperform decentralized information processors—its speed reducing even further the latent tension between theory and reality and its scale and resolution ensuring comprehensive scope and precision.[13] Aristotle's democratic answer may indeed have been more ethical than Plato's, but it won the battle of history because it was the more efficient.

This is not to say that AI's information-processing abilities must disrupt democracy itself, but it is a fact that the possibility of efficient centralization is poised to reshape the channels by which democracy is enacted. It might result in the instantiation of a democratic stock market, a single marketplace of ideas, assessed and valued at incredible speed. AIs may be able to incorporate the perspectives of billions or even trillions of future humans, thereby estimating accurately the current and distant future benefits and harms of—for instance—disruptive technological progress.

Put together, the collective imaginations of these minds could now be amalgamated into a single force, equalizing and potentially expanding beyond the Aristotelian methods. If a single mind could be engineered to represent the collective intelligence and values of a democratic polity—one immense philosopher to stand alongside the (elected) king—then the dual-headed vision of al-Farabi and Strauss could be realized.

Eventually, AI might close the gap between the sciences and the social sciences—political science being chief among the latter. The laws of human nature might become just as predictable, and policy just as reliable, as the laws of physics are to us now.

Unprecedented information processing will enable truly efficient centralization of policy by AIs. One might expect this to reinforce the perception of control by elites. However, the opacity of these systems—and the notion that their operation may be optimized in the absence of human interference—will work in the opposite direction. It is possible that, with time and experience, human control may come to seem less a necessity than a burden. Even as it might initially have felt terrifying for eighteenth-century European leaders to surrender control to the invisible forces of human self-interest, the political leaders of the twenty-first century may yet be required once more to humble themselves before a system that incorporates the wisdom of the masses in an entirely new form.

RULE BY REASON

If and when AI begins not only to process information for political purposes but also to make political decisions, new questions will pose an unfamiliar challenge to conventional political wisdom. Political science does not even provide guiding terms for such a shift. How might an onlooker assess the "superiority" of a strategic decision made by an AI, in the absence of interpretation of the logic by which the AI made the decision? The outcome, of course, would be one metric. But something significant would be lost, not least for historians, in the absence of a record setting out the principles of action.

In many future cases, humans might disagree with the plans produced by AI not because they are unwise but because the rationale for an AI's decision is beyond the scope of immediate human comprehension. It might be especially natural to wish to prevent or delay an outcome in which humans lose not only temporal control over machines' decision-making processes (the ability to intervene) but also logical comprehension of those processes (the ability to interpret), even after the fact.

In situations in which human and AI leadership diverge and disagree on matters of statecraft, which party's judgment should win out? Does the answer change if it becomes apparent that AI guidance, more than human counsel, effectively contemplates human end-states far in the future—and thus would be, in the long term, more beneficial than human discernment?

The utilitarian part of human intuition would be inclined to accept a farsighted AI's judgment, especially if the AI could explain the justification for its decisions. But, even then, situations may arise in which humans would protest a policy that, although ensuring the longevity of our species and of those persons not yet born, would be harmful to living humans here and now. Similarly, AIs might welcome rational outcomes, like settlements of military conflicts, that would be politically unacceptable to all parties to the conflict. The chances of outright rebellion would be high. Even if the concept of governance by machine were accepted, and even if that machine's logic were explicated, and even if that machine's decisions were rational, beneficial, and superior by some measures, we still might prove ungovernable. Among humans, only politics that incorporate an element of the intangible can be sustained. As Tolstoy writes, "If we admit that human life can be ruled by reason, the possibility of life itself is destroyed."[14]

Those nonrational elements of the human experience—history, above all, but also aesthetics, charisma, and emotional resonance—may hinder optimal outcomes, in some sense, but they are also foundational to our political groups. Rule by reason alone could dissolve nations. Overly rational AIs and their human partners could quickly lose control over or cause the disintegration of the structures of power in which they cooperate.

Then again, AI might be most valuable and most needed precisely where its logic appears strange, counter-intuitive, or just outright wrong. While AIs may be used to accelerate the resolution of problems toward known solutions—expanding human options by achieving a speed that avoids the political costs of delay—perhaps the even greater use of AI would be to think through what we cannot and land on entirely novel solutions. Indeed, this may be one of the key purposes of its creation.

But this attitude also carries risk, in proportion with its openness to opportunity. It leaves humans no basis on which to correct or ignore the potentially unacceptable and previously unimaginable decisions of an uninterpretable AI. The impulse not to hinder AI would only grow, exacerbated by the potentially evident superiority of AI governance as compared to past human governance.[15] An AI governor might deliver truly unbeatable results. If it did, stopping its use or circumscribing its scope would seem illogical—particularly in the context of geopolitical competition, in which refraining from use would seem to guarantee disadvantage.

Likewise, human leaders who become so accustomed to superior outcomes provided via AI would habitually depend on it for their own legitimacy. AIs might also develop their own biases: If a human leader loyal to an AI partner wished to overstay his or her term of appointment, would an AI intervene to halt such a violation of protocol?

PROMETHEUS

Throughout history, leaders who claimed to know what was best for their people, better than the people knew themselves, were soon disproved by reality.[16] Friedrich Hayek warned that central planning or similar forms of governance—including those forms not yet conceptualized during his time—would by definition forbid the expression of dissent.[17] Without, or even with, accurate information, planning can become a strong argument for prioritizing the collective over the individual—a pursuit of utilitarianism without obstruction. That is not inherently bad; but beyond a certain point, such a government can reach escape velocity, exiting the realm of normal governance and becoming eternal in ubiquity—with its subjects forced to be free, coerced for our own good. An AI-fueled administration could, in the words of one neuropsychologist, "claim to know what its people truly want and what will really make them happy...at best [used] to justify paternalism, at worst, totalitarianism."[18]

So long as we have known ourselves better than our king has known us, liberalism has been a restraining force. Now, as some have suggested, AI will "tell us who we are before we [know] ourselves," giving totalitarians not just an operational tool but a "philosophical weapon."[19] AIs could thus undermine the tenets of Immanuel Kant:

> No one has a right to compel me to be happy in the peculiar way in which he may think of the well-being of other men; but everyone is entitled to seek his own happiness in the way that seems to him best, if it does not infringe the liberty of others in striving after a similar end for themselves when their Liberty is capable of consisting with the Right of Liberty in all others according to possible universal laws.[20]

In the technical process of building AI, machines are taught that errors are wrong. That is purely useful insofar as it corrects humans' mistaken assumptions and suboptimal implementations. But another cause of error in our human systems is our individual free will—we are free to make the "wrong" choice. If an AI system decides to remove such errors, it faces two options: Remove us, or remove our free will. If free will is, by default, to be considered not a feature of intelligence but a bug, it may also be increasingly seen as a barrier preventing AI from accomplishing its own goals.

Humans have grown accustomed to both appreciating and despising the fact that the sheer strength of human will is what drives our world. History has been written in such a way as to show humans as the principal authors of our story; in that story, individual leaders have exercised the authority to see that empires rise and fall on their command or that great conquerors can exert enough stability to tame chaos. Total uncontrol may be rejected, even if it were known to maximize AI's ability to act in our (and future generations') interest.

The balance between control and uncontrol may be influenced by the problem of time discussed in the previous chapter, with humanity's and AI's perceptions operating on distinct timescales. AI is unlikely to be distressed by the speed at which it acts; it will view as natural the rate of change it creates. By contrast, the benefits to accrue from AI could be so dramatic—and, to humans, so unnaturally fast—that, sometime soon, our species may feel that we have drunk of it as much as we can tolerate and that any more would be more disorienting than pleasurable and therefore not worth the cost.

That does not necessarily spell the end of change. It simply means that, in order to bring along societies intact, the benefits that may be reaped from AI would need to be incorporated into human institutions incrementally, or at least more slowly than maximizers of distant future goods might like.

Many will be familiar, for example, with the story of Prometheus, the titan who stole fire from the gods to give to humans.[21] Fewer know that, in doing so, he was fully aware in advance of the cruel punishment he would receive from Zeus: namely, being chained to a rock for all eternity with an eagle eating his liver (which regenerated overnight) each day. What remains unanswered in the myth, however, is why Prometheus would choose to sacrifice himself, knowing in advance, as he also did, all the horrors that humanity would unleash with his gift. Perhaps he chose as he did, and is called a hero rather than a villain, because humanity would unleash very good things as well.

Cabined to partnership with humans, rather than unleashed toward domination of them, AI may help humans govern in obscure ways that are reminiscent and reflective of a similar kind of inaccessible forethought—"forethought" being the very meaning of the name *Prometheus* in Greek. But it may be accepted as a hero only if it exercises that forethought in service of humanity, a humanity that retains real individual agency and political coherence, even by way of some irrationality.

A NEW ACT

Examining the historical record, we might find most notable not the amount of change to be seen in politics but rather the remarkable absence of change. The same archetypes of leadership exist today as have existed for thousands of years: the tragic hero-prince, the treasonous adviser, the honorable lieutenant, the court jester, the shadowy puppet master, the unreliable mercenary.

If politics is theater, these familiar characters help us explain the otherwise inexplicable and animate the otherwise arcane.

Long before Catherine was great, Yaroslav wise, Ivan terrible, or Suleiman magnificent, they were not. And we admire these leaders in part because history has borne witness to their personal evolutions. We look with special awe upon those born in debilitating circumstances who through sheer conviction have escaped the shackles of adversity to rise to the top of their society. The Mamluk king Iltutmish, sold into slavery by his own brothers out of jealousy for his good looks and sharp intelligence, would serve Uzbek and Afghan slave traders in Bukhara and Ghazni before eventually being purchased by a Ghurid slave commander at a market in Delhi. Only two decades later, he, the slave of a slave, would rise through the ranks of the sultan's service, founding a "slave dynasty" atop the kingdom of his former masters.

Of course, the fall from power can be just as unexpected as the rise. No leader can guarantee immunity against being toppled by revolution or hanged by conspiracy. Whether by suicide like Nero, duel like Hamilton, or assassination like Gandhi, greatness cannot escape the graveyard.

As leaders in history, such figures hold a common stake in basic order and security. But as brothers and sisters in the family of man, they, like all of us, are not without jealousy, mutual suspicion, and sibling rivalry.[22] The same struggles are found in our own households as in ruling houses, in generations of common folk as in dynasties of royal reign. We can sympathize with the difficulty of overcoming jealousy and forging alliances with former enemies—as was done by Dowager Cixi, a former concubine who would come to rule China—just as we can despise the treason of General Mir Jafar, whose defection at the Battle of Plassey effectively handed control of India to the conquering British in return for the perfunctory title of "first Nawab of Bengal." If rulers were too godlike, too perfect, or too competent, they would not experience anxiety, despair, love, or envy. The familiar emotions and vulnerabilities encountered by all humans also turn the wheel of history.

Human politics is both commendable and condemnable because of its spiritual proximity to our own personal politics. Though Napoleon could tame Europe, he could not tame the heart of Joséphine; the rivalry between John Adams and Thomas Jefferson feels as familiar as our own brotherly squabbles; Tolstoy's *War and Peace* is no less about the principal events of history than about the individual lives that inhabit those events.[23] The men who in the fifteenth century CE would follow nineteen-year-old Joan of Arc to war were not so different from the men who in 324 BCE mutinied against Alexander the Great at Opis after he had conquered most of the known world. Alternating between the political and the personal, the real merges with fiction, historical chronicles with epic poems.

A mechanical politics would lack such narrative tension. Without the purging of rivals or the rapprochement with old enemies, the sudden rise to power or the rapid fall from grace, statecraft might become less relatable, less entertaining, even boring. With no more distinction between tragedy and comedy, the imperial court would become void of drama and intrigue.

Alternatively, the emergence of AI as a novel character could represent the inciting incident of a new act in human politics. It could change the natures of, and the dynamics among, the familiar human archetypes. But some things would not change; the finite lifespan of the human guarantees our narrative arcs, our rises, our falls. Our evolved social nature dictates our loves, ambitions, and moralities.

It is this same imperfect feature of our politics that we should aim to preserve and subsequently combine with the perfection of AI systems in becoming complementary partners. AI, lacking the inconsistencies of the human soul, will be unconstrained; for better and for worse, our flightiness has acted as a stifling cap on our potential but also as a reliable restraint on the extent of our evil.

Human rule today relies on our long experience with historical contingency. AI's value in governance, so far dormant, lies in its potentially perfect knowledge. Today's human leaders should prepare to be the first in a line of human sovereigns to face the struggle of locating a balance between leveraging the advantages—and, in some cases, the need—for AI in governance without going so far as to succumb to total dependency, instead finding the proper synthesis between the extremes of despotism and anarchy, merging the will of humans, the knowledge of machines, and the wisdom of history.

CHAPTER 5

SECURITY

FROM THE RECALIBRATION of military strategy to the reconstitution of diplomacy, AI will become a key determinant of order in the world. Immune to fear and favor, AI introduces a new possibility of objectivity in strategic decision-making. But that objectivity, harnessed by both the warfighter and the peacemaker, should preserve the subjectivity of human wisdom that is essential for the responsible exercise of force. More a revelation of our existing condition than a further discovery of the unknown,

AI in war will illuminate the best and worst expressions of our humanity. Even before the implementation of serious breakthroughs, an understanding will emerge of the extent to which AI can be, at once, the means of a conflict and the architect of its termination.

Humanity's long-standing struggle to constitute itself in ever-more complex arrangements, so that no state shall be in a position to have absolute mastery over others, has achieved the status of a continuous, uninterrupted law of nature. In a world where the major actors are still human—even if equipped with AI to inform, consult, and advise them—we should still enjoy a degree of stability based on shared norms of conduct among the actors involved, subject to the tunings and adjustments of time.

But if AI emerges as a practically independent political, diplomatic, and military set of entities, that would force the exchange of the age-old balance of power for a new, uncharted disequilibrium. The international concert of nation-states—a tenuous and internally shifting equilibrium achieved in the last few centuries—has held in part because of the inherent equality of the players. A world of severe asymmetry—for instance, if some states adopted AI into political leadership more readily than others—would be far less predictable. In cases where some humans might face off militarily or diplomatically against a highly AI-enabled state, or against AI itself, humans could seem defenseless to survive, much less compete. Such an intermediate order could witness an internal implosion of societies and an uncontrollable explosion of external conflicts.

Beyond a minimum of security, humans have long entered into combat in pursuit of triumph or in defense of honor. But machines—for now—lack any real conception of either triumph or honor. They might prosecute a kind of war unique in human experience. But what choices would AIs make? Might they never go to war, choosing instead (for instance) immediate, carefully divided transfers of territory based on complex calculations of relative strategic advantage? Or might they—prizing an outcome and deprioritizing individual lives—spiral into wars of human attrition? In one scenario, our species could emerge so transformed as to avoid entirely the brutality of human conduct. In another, we would become so subjugated as to return us to a barbaric past.

Is there no other way? Thus far, neither the patient overtures of diplomacy nor the shocking horrors of war have ever succeeded in permanently encoding in humans—let alone another species—an aversion to destruction. Is it nevertheless possible, thanks to the age of AI, that conditions will finally materialize for a perpetual peace to emerge at last?

ESPIONAGE AND SABOTAGE

Even as nations search for ways to keep the technology safe, they are simultaneously fixated on how to "win the AI race."[1] In part, that response is understandable. Culture, history, communication, and perception have conspired to create among today's primary states a diplomatic situation that fosters insecurity and suspicion on all sides. AI, as a dominant feature of the global future, forms a low flashpoint in an already volatile combination in which each side believes that an incremental tactical advantage could be decisive for its long-term benefit.

If, following from the instinct for self-preservation, each human society wishes to maximize its unilateral position, then the conditions would be set for a psychological contest between rival military forces and intelligence agencies, the likes of which humanity has never faced before. Today, in the years, months, weeks, and days leading up to the arrival of the first superintelligence, a security dilemma of existential nature awaits. The logical first wish for any human actor coming into possession of such a capability might be to attempt a guarantee of singular continued existence. Any such actor might also reasonably assume by default that its rival, under the same uncertainties and facing the same stakes, would be pondering a similar move.

Even if a dominant nation were to stop short of war, a superintelligent AI could subvert, undermine, and block a competing program. For instance, AI promises both to strengthen conventional viruses with unprecedented potency and to disguise them with similar thoroughness. Like the computer worm Stuxnet—the cyber-weapon said to have ruined a fifth of Tehran's uranium

centrifuges before detection—an AI agent could sabotage a rival's progress in ways that obfuscate its presence,
thereby leading enemy scientists down futile pathways of
inquiry.[2]

With its unique capacity for manipulation of weak-
nesses in human psychology *en masse*, an AI could also
hijack a rival nation's media, producing a deluge of synthetic disinformation so alarming as to inspire mass
opposition against further progress in that country's AI
capacities. Or, to another nation's chief AI scientist, it
might target and personalize communications to render his or her perception of personal grief so great as to
degrade the capacity for effective leadership.

Assessing the state of the competition will be even
more challenging. Already the largest AI models are being
trained on secure networks disconnected from the rest of
the internet. Some executives believe that AI development
will itself sooner or later migrate to impenetrable bunkers
whose supercomputers are powered with nuclear reactors.[3] Data centers are even now being built on the bottom of the ocean floor.[4] Soon they could be sequestered
in orbits around Earth. Corporations or nations might
increasingly "go dark," ceasing to publish AI research so
as not only to avoid enabling malicious actors (as publicly
stipulated) but also to obscure their own pace of construction (as privately intended). To distort the true picture of
their progress, others might even try deliberately publishing misleading research, with AI assisting in the creation
of convincing fabrications.

There is a precedent for such scientific subterfuge: In 1942, the Soviet physicist Georgy Flyorov correctly inferred that the U.S. was building a nuclear bomb after he noticed that the Americans and the British had suddenly stopped publishing scientific papers on atomic fission.[5] Today, however, such a contest would be made all the more unpredictable given the complexity and ambiguity of measuring progress toward something so abstract as intelligence. Although some see "advantage" as commensurate with the size of the AI models in their possession, a larger model is not necessarily superior across all contexts and may not always prevail over smaller, inferior models deployed at scale. Smaller and more specialized AI machines might operate like a swarm of drones against an aircraft carrier—unable to destroy it, but sufficient to neutralize much of its capacity.

To some actors, an overall advantage would be signaled by the achievement of a given capability. The problem with this line of thinking is that *AI* refers merely to a process of machine learning that is embedded not just in a single technology but in a broad spectrum of technologies. Capability in any one area may thus be driven by factors entirely different from capability in another. In these senses, any "advantage" as ordinarily calculated may be illusory.

Moreover, as demonstrated by the exponential and unforeseen explosion of AI capability in recent years, the trajectory of progress is neither linear nor predictable. Looking ahead, experts continue to disagree about the development of superintelligence: Is it merely a matter of scaling and the implementation of existing learning infrastructures, or will superintelligence require additional creative and scientific innovations?[6] Conceivably, the transition from narrow intelligence to general intelligence and then to superintelligence could pass without obvious signs of evolution—particularly if humans do not develop a uniform conception of what to look for. Even if one actor could be said to "lead" another by an approximate number of years or months, a sudden technical or theoretical breakthrough in a key area at a critical moment could invert the positions of all players.

In such a world, where no leaders could trust their most solid intelligence, their most primal instincts, or even the basis of reality itself, no real contender for superintelligence could be blamed for acting from a position of maximum paranoia and suspicion. Leaders are no doubt already making decisions under the assumption that their endeavors are under surveillance or harbor distortions created by malign influence. Defaulting to worst-case scenarios, the strategic calculus of any actor at the frontier would be to prioritize speed and secrecy over safety. With human perception unable to detect or defend against AI-enabled attempts at suppression, human leaders could be gripped by the fear that there is no such thing as second place. Under pressure, they might prematurely accelerate the deployment of AI as deterrence against external disruption.

Today, we are still behind an involuntary veil of igno-rance. The ultimate winners (if winning can be defined) of the competition for AI dominance are as yet unknown. Every enterprise in pursuit of the front rank is a potential rival. This uncertainty will produce instability.

Traditionally, after the emergence of a new power, a bloody contention has been necessary before rivals can settle into a new status quo that is at least minimally acceptable to all parties. However, in a world of nuclear weapons complicated by such a rapid reconfiguration by AI, there may be no opportunity to validate a new con-sensus by the established principles and instruments of warfare.

In the event that the identity of a winner does crys-tallize, mere competition could devolve into conflicts driven by desperation and fear. In such an instance, cer-titude could be even more destabilizing than precaution. Put together, perfect deterrence with unlimited speed and maximum accuracy translates to total victory. Historical pursuits of a monopoly on force have typically caused an attendant expansion in the assembly of dark designs by others. When such capacities are considered in the con-text of a precariously balanced world, some states may deem the advent of AI threatening enough to demand a nuclear response. Having obviated conventional wars, will AI then drag us backward into nuclear ones?

In previous arms races, evolutionary instinct mixed with competition among inventors has enabled the

creation and operation of instruments designed to preserve homeland societies. Here, too, defensive AI systems
could neutralize attacks by adversaries, for instance by
updating currently vulnerable software and other systems, or, if tasked with the surveillance of competing
programs, by serving as an early warning system. However, in this case, new threats—an AI-enabled bioweapon,
a sudden nuclear proliferation, or even a misaligned AI
itself—might emerge so quickly and silently, and wreak
such devastation, as to preempt any response.

Though we may not yet have entered such a moment,
preparations should be made in advance to manage the
existential competition of the AI age and its attendant
risks. A determined or desperate actor with secondary but sizable strength will be vigilant in the watch for
superior AIs. If that secondary actor perceives—even
mistakenly—that the dominant other is within grasp of a
totalizing capability, it may launch preemptive electronic
or physical strikes, setting off a cycle of unimaginable
escalation and retaliation and the spiraling possibility of
mutual destruction.

A glimmer of hope—at least for today—lies in our
uncertainty. Ambiguity can be a fertile ground in which
to cultivate dialogue. Unsure of how any one nation or
group might be advantaged or disadvantaged by AI decisions made today, global leaders have a window of opportunity to engage in discussions on the basis of humanity's
collective survival.

MANAGING EMERGENCE

If the future is a competition to reach a single, perfect, unquestionably dominant intelligence, then it seems likely that humanity will either lose control of an existential race among multiple actors or suffer the exercise of supreme hegemony by a victor unharnessed by traditional checks and balances. The closer the margin by which the race is won, the higher the probability of human miscalculation.

Unipolarity may be one pathway that could minimize the risk of extinction. Were the existing front-runner able to extend its lead to a threshold where no other entity believed it possible to close the gap, there might be greater certainty of ensuring some level of stability. At least for a while, the basis of world order—the ceaseless pursuit of transitory and fragile equilibria among balanced forces—might no longer be desirable.

Alternatively, it might be possible—though contingent on the suppression of natural human instincts—to broker an agreement among competing entities to assure an agreed-upon period of cooperation or even to merge into a single enterprise the multiple competing efforts at the head of the line. But achieving unification across geopolitical and commercial hostilities would require an extraordinary amount of courage and foresight. Any actor at or near the front would expect to be included in the company of victory. Within that prime category, indecisive or still-lagging actors would naturally achieve the greatest benefit from entering into the arrangement, thereby becoming well placed to gain parity and possibly to pull ahead. By contrast, other leading actors might find the sacrifice of their pole position intolerable, and their trust in altruism might prove too brittle to withstand the temptations of outright defection.

A similar option would be a negotiation to distribute and delegate power. When confident that victory is imminent, the best-positioned actor might attempt to persuade its rivals to surrender by guaranteeing their privileged access to superintelligent AI. But never has trust in the human commitment to one's fellow man, much less to one's former adversary, endured forever. If humans were to manage, against all odds, to forge such an agreement, it is far from clear how its implementation would be governed.

Indeed, simply collecting the sharpest peaks on the "island" of knowledge might exacerbate the dynamics that their consolidation was designed to prevent in the first place. Human history has yet to produce any record of such an attempt, much less a success. Moreover, such an arrangement would require a dramatic reorientation of diplomatic strategy. Historically, the very basis of world order has been maintained by the ceaseless pursuit of a delicate equilibrium of balanced forces. Here, by contrast, nation-states would be pursuing a hegemonic stasis largely unfamiliar to human practice. The dominant bloc of states likely would be the original creators of AI systems and the suppliers of their most critical components, as well as the producers of the talent to develop and refine them. The nondominant majority might be relegated to the status of tributaries, supplying data and other goods and repaid with some level of access to AI's discoveries, governance systems, and defenses.

In any case, this scenario is neither our preference nor our prediction. Rather, we believe there will be not just one supreme AI but rather multiple instantiations of superior intelligence in the world. In that case, a different set of possible futures emerges. Our strongest creations, acting as countervailing forces, could be better equipped than humans to exert and maintain an equilibrium in global affairs inspired (but not constrained) by human precedent. Nonhuman intelligence could thus manage its own emergence, at least in the realms of national security and geopolitics.

But could AI engineer a sustainable division of dominion? Possibly, yes. Diplomatic negotiations, underneath their layers of human emotion and psychology, are at their core a species of applied game theory, which is itself a branch of mathematics. Even if the practice of diplomacy originated as a form of art (albeit one exclusive to the realm of human conduct), it might increasingly morph into a science. As it does, it could transcend the rather mixed track record of human diplomacy when it comes to identifying and pursuing opportunities for compromise. Already, early AI models in the West have shown great promise in their ability to apply strategy, at least in game-play, while China has moved a step further, leveraging machine intelligence to discharge the duties of human diplomats.[7]

The human tradition of diplomacy began with the simple need to convey messages between societies safely and reliably. Gradually, envoys from afar would come to enjoy special treatment, while states violating the new norms would seldom escape sanction. When representatives of Xerxes demanded symbols of submission from the Greek city-states, the Persian diplomats were thrown into pits and down wells. Herodotus records that the gods would later punish Sparta for this deed.[8] About seventeen centuries later, when a senior Mongol diplomat was executed on the order of a Persian shah, Genghis Khan commanded his fearsome horde to destroy the entire Khwarezmid empire in retribution.[9] Still, keeping open the channels of communication, even—and especially—in times of war, became custom; over time, a basic consensus formed that it might be better to hear the message than to kill the messenger.

Were AIs to conduct diplomacy among themselves, they might be intentionally trained to accept the same custom, or they might incidentally develop a bias in favor of the additional information that these customs guaranteed. Surely there would also be aberrations. While perhaps no heads would roll, machines might find an equivalent way of pursuing their narrow interests. Here their total rationality, if humans could accept it, might increase the probability of at least a safer start.

The provision of machine solutions to human problems as fundamental as diplomacy and security will, however, lead naturally to further dependence on AI's abilities—a dependence difficult to break if and when decisive human intervention in international affairs becomes necessary. At least human statecraft, however imperfectly applied, enforced upon us the responsibility for our own choices. Choosing to depend on AI, by contrast, may reduce our ability to trust the basic human judgment on which we could reliably subsist in our more predictable but primitive past.

Why take such a risk? For one thing, to avoid the dilemma of mutually destructive competition or a hegemony produced by emergent superintelligence, but also to protect ourselves from other threats looming on the horizon. With every passing year, with each new technological breakthrough, the minimum threshold for destroying all of human society has lowered.

Nordic mythology tells the story of Baldur, son of the god Odin and Queen Frigg, alarmed by prophecies of their son's impending gruesome death.[10] Queen Frigg, determined to shield him from whatever fate lay ahead, traveled tirelessly across the nine realms of Earth, casting powerful magic on every animal, element, plant, and plague so that none could be used as a weapon against him. But the trickster god Loki, disguised as an old woman, coaxed from the queen the information that her divine cloak of protection covered everything on Earth except mistletoe, the most harmless weed of all. At a feast held to celebrate

Baldur's immunity, the gods, to showcase the queen's fine work, took turns launching at him weapons of every make. Loki, however, compelled his blind brother Hodr to shoot an arrow tipped with mistletoe, piercing Baldur's chest and killing him by means of the only object in Valhalla exempt from the queen's otherwise comprehensive guardianship.

The meaning of this ancient myth for our modern predicament is simple and chilling: As threats grow ever more obfuscated and sophisticated, humanity's defenses against them must be ever more perfect, since the slightest mistake or omission could spell defeat. And to achieve that level of perfection, we might well need the assistance of AI.

The exercise is thus to decide which risk is lower and therefore should come first: surviving the development of AI, surviving other parallel revolutions like the advent of synthetic biology, or surviving potential catastrophes such as radical climate change. Granted, smaller AIs may assist in the reckless invention of existential technologies; even if the creators of these new means were themselves perfectly and properly regulated, others would likely be less careful and more harmful. But larger AIs could assist in defending against the same technologies, enabling actual perfection in tactical defensive decision-making.

For instance, if developed, individualized biodefense— whereby AI-enabled nanobots would be deployed in our bloodstream to remove anything that failed to match a recognized biosignature—would be a more agile corrective than our previous response to biological threat. Similarly, AI may generate new materials and new processes to reduce carbon emissions, lowering the risk of climate catastrophes.

No doubt, it is a risk for AI to assume early and sustained responsibility for the species and societies behind its own conception, but the traditional pathways, which require perfection in human performance, may be even riskier. Best, in our current view, would be to have AI working before and not after humanity has to confront the proliferation of new threats to survival.[11] The appropriate question under this assumption is this: How can humans accelerate only desirable pathways for AI, while delaying the undesirable?

AI is an unselectively destabilizing force; its emergence, if unmanaged, poses as great a risk to its creators as to its users. This is precisely what might compel initially unwilling rivals to consider otherwise implausible agreements. We believe that in diplomacy, defense, and perhaps elsewhere, some of the risks of AI can be managed successfully only by AI itself. Pandora's box has already been opened; even if it had not been, the benefits of AI still appear to outweigh the risks.

It is our contention, therefore, that our "vulnerable world" (to borrow a phrase from Nick Bostrom) may well require intervention by AI in order to survive some of the challenges and threats of AI's technological development.[12] The question remains: How must humans behave when facing a future that simultaneously demands and forbids our continued control?

A NEW PARADIGM OF WAR

For almost all of human history, war has been fought in a defined space in which one could know with reasonable certainty the capability and position of hostile enemy forces. The combination of these two attributes offered each side a sense of psychological security and common consensus, allowing for the informed restraint of lethality. Only when enlightened leaders were unified in their basic understanding of how a war *might* be fought could opposing forces control *whether* a war should be fought.

Speed and mobility have been among the most predictable factors underpinning the capability of any given piece of military equipment. An early illustration is the development of the cannon. For a millennium after their construction, the Theodosian Walls protected the great city of Constantinople from outside invaders. Then, in 1452, an artillery engineer from the kingdom of Hungary—at that point under Byzantine suzerainty— proposed to Emperor Constantine XI the construction of the Basilic, a giant cannon that, firing from behind the defensive walls, would pulverize outside intruders. But the complacent emperor, possessing neither the material means nor the foresight to recognize the technology's significance, dismissed the proposal.

Unfortunately for him, the Hungarian engineer turned out to be a political mercenary: Switching tactics (and sides), he updated his design to be more mobile—transportable by no fewer than 60 oxen and 400 men—and approached the emperor's rival, Turkish Ottoman sultan Mehmed II, who was preparing to besiege the impermeable fortress. Winning the young sultan's interest with his claim that this gun could "shatter the walls of Babylon itself," the entrepreneurial Hungarian helped the Turkish forces to breach the ancient walls in only 55 days.[13]

The contours of this fifteenth-century drama can be seen again and again throughout history. In the nineteenth century, speed and mobility transformed the fortunes first of France, as Napoleon's Grande Armée overwhelmed Europe, and then of Prussia, under the direction of Helmuth von Moltke (the Elder) and Albrecht von Roon, who capitalized on the newly developed railways and embraced distributed control to enable faster and more flexible maneuvering. Similarly, blitzkrieg—an evolution of the same German military principles—would be used against the Allies in World War II to great and terrible effect.

"Lightning war" has taken on new meaning—and ubiquity—in the era of digital warfare. Speed is instantaneous. Attackers need not sacrifice lethality to sustain mobility, as geography is no longer a constraint. Although that combination has largely favored the offense in digital attacks, an AI era could see the increase of the velocity of response and re-enable cyber defenses on par with cyber offenses.

In kinetic warfare, AI will provoke another leap forward. Drones, for instance, will be extremely quick and unimaginably mobile. Once AI is deployed not only to guide one drone but to man fleets of them, clouds of drones will form and fly in sync as a single cohesive collective, perfect in their synchronicity. Future drone swarms will dissolve and reconstitute themselves effortlessly in units of every size, much as elite special-operations forces are built from scalable detachments, each of which is capable of sovereign command.

In addition, AI will provide similarly speedy and flexible defenses. Drone fleets are impractical if not impossible to shoot down with conventional projectiles. But AI-enabled guns firing rounds of photons and electrons (instead of ammunition) could re-create the same lethal disabling capacities as a solar storm that can fry the circuitry of exposed satellites. Once more, speed and mobility will be beyond human capacity—and potentially equalized on defense and offense.

With speed and mobility no longer defining variables, the gap in capability between competing entities would now rely on precision, immediate impact, and strategic application.

AI-enabled weapons will be unprecedentedly exact. Limits to the knowledge of an antagonist's geography have long constrained the capabilities and intentions of any party to hostilities. But the alliance between science and war has come to ensure increasing accuracy in our instruments, and AI can be expected to make another breakthrough—or many. AIs will thus shrink the gap between original intent and ultimate outcome, including in the application of lethal force. Whether land-based drone swarms, machine corps deployed in the sea, or possibly interstellar fleets, machines will possess highly precise capabilities of killing humans with little degree of uncertainty and with limitless impact. The bounds of the potential destruction will hinge only on the will, and the restraint, of human and machine.

That being so, the AI age of warfare will be reduced primarily to an assessment not of an adversary's capabilities but rather of its intentions and their strategic applications. In the nuclear age, we have already entered such a phase, in a sense—but its dynamics and significance will come into much sharper focus as AI proves its worth as a weapon of war. Thus, the key question is this: What will AI-enabled commanders want and need?

With such valuable technology at stake, humans are not likely to be considered the primary targets of AI-enabled war. AIs could in fact remove humans as a proxy in warfare entirely, making war less deadly but potentially no less decisive. Similarly, territory alone seems unlikely to provoke AI aggression—but data centers and other critical digital infrastructure certainly could. (Supercomputers are likely to be hidden and storage of intelligences distributed to secure a stronger chance of operational continuity and to defend against a mechanical "decapitation strike.")[14]

Surrender, then, will come not when the opponent's numbers are diminished and its armory empty but when the survivors' shield of silicon is rendered incapable of saving its technological assets, and finally its human deputies. War could evolve into a game of purely mechanical fatalities, the deciding factor being the psychological strength of the human (or AI) who must contest to risk, or forfeit to prevent, a breakthrough moment of total destruction.

Even the motives governing the new battlefield would be alien, to some extent. G. K. Chesterton tells us "the true soldier fights not because he hates what is in front of him, but because he loves what is behind him."[15] An AI war is unlikely to involve love or hate, let alone a concept of soldierly bravery. On the other hand, it may still incorporate ego, identity, and loyalty—though the nature of those identities and loyalties may not be consistent with today's.

The calculation in warfare has always been relatively straightforward: Whichever side first finds intolerable the excruciations caused by its antagonist's might becomes subject to conquest and conversion, and only then, perhaps, to negotiation. The consciousness of one's own shortcomings reliably and naturally has produced restraint. Without such awareness, and with no sense for (and thus with total tolerance of) pain, one cannot but wonder what, if anything, would prompt restraint in an

AI that has been introduced into warfare, and what would conclude the conflicts it wages. Would a chess-playing AI, if it had never been informed of the rules dictating the end of the game, have played to the last pawn?

GEOPOLITICAL RESTRUCTURING

In every age of humanity, almost as if in obedience to some natural law, there has emerged, as one of us once put it, a unit "with the power, the will, and the intellectual and moral impetus to shape the entire international system in accordance with its own values."[16] Subsequent to that entity's appearance, other units become linked in novel arrangements, building unpredictable dependencies in times of crisis and constantly threatening to undo geopolitical balances of power. In some cases, the resultant system upends existing authorities; in others, it entrenches them.

The most familiar arrangement of human civilizations is that of the Westphalian system as conventionally understood. The idea of the sovereign nation-state, however, is only a few centuries old, having emerged from treaties that are collectively known as the Peace of Westphalia in the mid-seventeenth century. It is not the preordained unit of social organization, and it may not be suited for the age of AI. Indeed, as mass disinformation and automated discrimination trigger a loss of faith in that arrangement, AI may pose an inherent challenge to the power of national governments. Compounding the issue would likely be the psychological disorientation and possible retreat from reality detailed earlier in this book. Alternatively, AI may well reset the relative positions of competitors within today's system. If its powers are harnessed primarily by nation-states themselves, humanity could be forced toward a hegemonic stasis, or else toward a new equilibrium of AI-empowered nation-states. But it could also be the catalyst of an even more fundamental transition—a shift to an entirely new system, in which state governments would in turn be forced to abandon their central role in the global political infrastructure.

One possibility is that the companies that own and develop AI will accrue totalizing social, economic, military, and political power. Today's governments are forced to contend with their difficult position as hosts and cheerleaders for private corporations—lending their military power, diplomatic capital, and economic heft to promote their homegrown interest—and with their simultaneous role as supporters of the everyman who harbors suspicions of monopolistic greed and secrecy. That may prove an untenable contradiction. And, as detailed earlier in this book, the emergence of AI will make governance by any established institution more difficult.

Meanwhile, corporations could form alliances to consolidate their already considerable strength. Those alliances might be built on complementary advantages and the profit of amalgamation or, alternatively, on a shared philosophy of development and deployment of AI systems.

These corporate alliances might take on traditional nation-state functions, though rather than seeking to define and expand bounded territories, they would cultivate diffuse digital networks as their domains.

And there is still another alternative. Uncontrolled, open-source diffusion could give rise to smaller gangs or tribes with substandard but substantial AI capacity, sufficient to administer to, provide for, and defend themselves within some limited scope. Among human groups that reject established authority in favor of decentralized finance, communication, and governance, proto-anarchy could win out. Or such groupings might incorporate a religious dimension, perhaps driven by one of the conceptions of AI and divinity explored in an earlier chapter. After all, in terms of reach, Christianity, Islam, and Hinduism have all been larger and longer-lasting than any state in history. In the age to come, religious denomination, more than national citizenship, might conceivably prove the more relevant framework for identity and loyalty.

In either future, whether dominated by corporate alliances or diffused into loose religious groupings, the new "territory" that each group would claim—and over which they would fight—would not be inches of land but perhaps digital devices, as a signal of the loyalties of individual users. Linkages among these users and any administration—undoubtedly affected by the complex effect of AI on the status of traditional centralized government—would subvert the traditional notion of citizenship, and agreements among the entities would be unlike ordinary alliances.

Historically, alliances have been forged by individual leaders and have served to augment a nation's strength in case of war. By contrast, the prospect of citizenships and alliances—and perhaps conquests or crusades— structured around the opinions, beliefs, and subjective identities of ordinary people in times of peace would require a new (or very old) conception of empire. It would also force a reassessment of the obligations entailed in pledging allegiance and the cost of exit options, if indeed any were to exist in our AI-entangled future.

PEACE AND POWER

The foreign policies of nation-states, and thus international systems, have been built and then adjusted by balancing idealism and realism. The temporary balances struck by our leaders are seen in retrospect not as end states but as only ephemeral (if necessary) strategies for their time. With each new age, this tension has produced a different expression of what constitutes political order. A leader cannot merely realize an option that falls along an existing and already-considered spectrum. Instead, governors must make at least some choices that derive (or appear to derive) from inspiration—frequently encouraging the pursuit of objectives that lie beyond the reach of practical attainment.

The dichotomy between the pursuit of interests and the pursuit of values—or between a particular nation-state's advantage and the global good—has been part of this unending evolution. In the conduct of their diplomacy, leaders of smaller states historically have responded straightforwardly, prioritizing the necessities of their own survival. By contrast, those responsible for global empires, with the means to realize additional goals, have faced a more agonizing predicament.

Since the beginning of civilization, as human units of organization have grown, they have simultaneously achieved new levels of cooperation. But today, perhaps due to the scale of our planetary challenges as well as to the material inequalities evident among and within states, a backlash against this trend has surfaced. Might AIs prove themselves commensurate with this still-grander scale of human governance, capable of seeing with granularity and fidelity the interplay of the globe and not merely the imperatives of the nation? Could they be relied upon to calculate—more precisely than we have ever done before—first our interests and our values, and then their correct proportion and relation to each other?

It would be unrealistic to expect, as one of us has put it before, that human leaders will reliably "confine our actions to situations in which our moral, legal, and military positions are completely in harmony and where legitimacy is most in accord with the requirements of survival."[17] For humans, this is still true. Yet we harbor a hope that AIs, deployed for political ends at home and abroad, might do more than just illuminate balanced trade-offs. Ideally, they could provide new, globally optimal solutions, acting on a longer time horizon and with higher resolution than humans are capable of, and thus bringing each of our competing human interests into alignment. In the coming world, machine intelligences navigating conflict and negotiating peace might help clarify, or even surmount, our traditional dilemmas.

However, if AI were indeed to fix problems that we should have hoped to solve ourselves, we could face a crisis of confidence: that is, overconfidence on the part of some and lack of confidence on the part of others. To the former, once we understand the limits of our own ability for self-correction, it may be difficult to admit that we have come to cede too much power to the assumed wisdom of machines in handling existential issues of human conduct. To the latter, the realization that simply removing human agency from the handling of our affairs has been enough to solve our most intractable problems might reveal too explicitly the shortcomings of human design. If peace has always been but a simple voluntary choice, the price of human imperfection has been paid in the coin of perpetual war. To know that a solution has always existed but has never been conceived by us would be crushing to human pride.

This is one especially poignant instance of the dilemma of dependence—and subsequent perceived inferiority—explored in an earlier chapter. But, in the case of our security, unlike that of our displacement in scientific or other academic endeavors, we may more readily accept the impartiality of a mechanical third party as necessarily superior to the self-interestedness of a human—just as humans easily recognize the need for a mediator in a contentious divorce. It is our belief, and hope, that in this case some of our worst traits will enable us to exhibit some of our best: that the human instinct toward self-interest, including at the expense of others, may prepare us for accepting AI's transcendence of the same.

CHAPTER 6

PROSPERITY

THE FINNISH NATIONAL epic *Kalevala* begins with Väinämöinen, the first man to bring trees and life to a previously barren world, now washed up on the shores of distant Pohjola after an exhausting defeat in a battle at sea.[1] After nursing the hero back to health, Louhi, the wicked queen who rules over Pohjola's dark and shadowy land, demands repayment in return for his release. Not content with mere gold or silver, the old hag of the north requests what then existed only in myth: the Sampo, a

137

magical machine capable of producing for its owner an endless fountain of wealth.

With his brother Ilmarinen—architect of heaven's dome and the only person capable of creating anything like the Sampo—Väinämöinen strikes a deal with the powerful witch, promising that if she releases him, he will honor his debts by sending his brother back in his place. Louhi then seduces Ilmarinen with the prospect of receiving one of her beautiful maiden daughters as a bride, and the master craftsman enthusiastically does as Louhi commands.

Summoning divine winds to work the bellows for three days, the eternal hammerer forges the finest materials in the realm—"the tips of white swan feathers, the milk of greatest virtue, a single grain of barley, and the finest wool of lambskins"—into the Sampo, grafting spigots onto its sides that gush forth an infinite supply of grain, salt, and coin.[2] But just as Ilmarinen pulls the wealth-making machine from the forge's flames, Louhi gleefully snatches it out of his hands and locks it inside a mountain vault. Henceforth, Pohjola will prosper from its limitless productive powers while Ilmarinen is left bitter and dejected.

Many years later, Ilmarinen and Väinämöinen return together to correct the witch's injustice. Upon arrival in the kingdom of plenty, they threaten to take the Sampo by force if they do not receive half of its profits. A fierce battle ensues at sea, and amid the chaos the Sampo sinks deep into the black depths where it remains, beyond retrieval, churning out riches and commodities without a master while rendering the ocean's water salty until this very day.

Similar stories describing machines of bounty exist throughout the world: the Akshaya Patra, a bottomless copper vessel described in the Hindu epic *Mahabharata*; the magical Cauldron of Plenty in the Irish myth of the god Dagda; the *Uchide-no-kozuchi*, a magic mallet in Japanese lore capable of "tapping out" anything upon command, including houses, clothing, and even humans.[3]

Those building AI today believe that their creation will constitute this fully stocked granary, this magic mill, this cornucopia overflowing with flowers, fruit, and corn. As mythology warns, however, creation alone will not be enough. To fulfill its potential, AI must, in its development and deployment, be aided by and coupled with suitable institutional changes and wise policy design. AI should be used to loosen, and ideally remove entirely, the bonds of servitude that in the past have dominated humanity's social and economic relations, and to launch us into a future characterized by less poverty and inequality.

This goal is extraordinarily ambitious by any measure, and one can rightly raise questions. Still, what if AI were indeed to offer a bridge to a new golden age? Even a partial success could amount to a civilizational renaissance.[4]

GROWTH AND INCLUSIVITY

In March 2016, after losses in three straight games, Lee Sedol, the Korean grand master of the Chinese game Go, felt neither anger nor sadness but wonder. He never imagined that he—having dedicated his entire life to the mastery of this ancient game—could lose to AI, an ultramodern foe. And yet, only a game earlier, his computational opponent, known as AlphaGo, had played a move—number 37—so unorthodox as to force him to consider anew that machines might not only possess raw capability but also wield powers of creativity.

By now, awed by his extraordinary competitor, Lee was no longer playing solely in pursuit of victory—which he had already conceded in their best-of-five match— but was instead striving for a beautiful finish. In the next and fourth game, he responded to Move 37 with Move 78: a stroke of corresponding artistry that propelled him against all odds to victory in what is still the only game that AlphaGo has ever lost. In Korea and around the world, commiseration gave way, temporarily, to celebration.

Throughout that week in Seoul, the world's number one player, with only the occasional reprieve for a smoke on the outdoor patio of the sponsoring luxury hotel, had fought a lonely fight, locked in a contest with a unique adversary that he never expected to face while playing on behalf of a team—humanity—that he had never sought to represent. Their contest will be most remembered not for its final outcome but, all things considered, for the astounding capacity on display by human beings: namely, both Lee himself and those who developed his machine rival.[5]

The motto of DeepMind, the company behind the victory in Korea, is "First, solve intelligence; then, use intelligence to solve everything else."[6] Intelligence, as an engine of fresh creation, is set to change our understanding of everything. Facing this great unknown, we could become overwhelmed. But in at least some contexts, it may be wise to respond with the attitude shown by Lee Sedol, who in facing AI treated it as an inspiration rather than as a rival.

Lee Sedol's situation was unique—that of an expert at the top of his field, playing against an AI as an experiment. As a result, he may have been naturally suited to a posture of awe and imagination, as opposed to one of resentment. Many other people will respond to AI much more negatively.

Refraining from rivalry will be particularly difficult in what appear to be zero-sum situations, perhaps especially AI's potential displacement of human labor. In this chapter we attempt to address that supposed zero-sum dynamic—which we see as largely misperceived— and to describe what we believe could be abundant and widely distributed benefits to humanity, even, potentially, in a world without work.

For most human beings throughout most of history,

labor has not been a game one hopes to win or an art form one hopes to master but instead an unfulfilling and brutal burden, enforced through social structures that keep workers bound in service. Though these structures may have helped to maintain stability, they have invariably served also to torment the human spirit.

In the *Bhagavad Gita*, part of the *Mahabharata*, a dialogue between the warrior prince Arjuna and his charioteer—the deity Krishna in disguise—offers a pertinent discourse on the socioreligious hierarchies that long provided order to Indian society. As Arjuna hesitates on the battlefield to raise a sword against his own kin, Krishna explains to him in no ambiguous terms that there can be no deviation from duty, no divergence from destiny: "Better botch your job than gain perfection in your neighbor's; die if you must, but do not run the risk of alien labors."[7]

In this context, everyone plays a specific role, no matter how unfulfilling, as unfairly determined by birth. Thus, according to the *Gita*, "The duty of Brahmins [the highest caste] is to be peaceful and wise; of soldiers, to fight; of the middle classes, to care for farming and trade; and of serfs, to perform menial services."[8] Only if everyone faithfully discharges the duties of his or her station can a society succeed. Those who do so in this life have a chance at higher status in the next; for those who don't, suffering in the next reincarnation awaits.

Hindu castes, of course, were not alone. Aristotle's

political theory involved strict social roles and duties. Slavery—underpinned by law, force, and psychological torment—became in parts of the world the primary cruel institution for extracting labor and enforcing constructed social rank.

Over the last two centuries, capitalist democracies have largely replaced both caste and captivity with markets of meritocracy; churchmen have elevated and scholars have documented the societal values of a strong work ethic; and workers have acquired the arts of bargaining and strikes. But even so, and whether our human labors have been employed by the divine, or by government, or to guarantee a wage, our exertion of mind and body has generally been not so much for ourselves as in service to another.

Many a war has been caused by or else has resulted in changes to the question "Who gets what—and why?" (to borrow the title of the 2015 bestseller by the economist Alvin Roth). Earth's relatively fixed supply of land, labor, and capital has ensured that scarcity—not abundance—has been the predominant paradigm of economic theory and practice. Fierce battles are fought over how to divide what has been created, and even more often over how to distribute what little remains. These frictions pertain as much within societies as among them, even in peacetime, with citizens debating the roots of relative disadvantage and calling for redistribution to address the fact of widespread suffering.

An enlargement of the total amount of wealth available for redistribution, and subsequently an actually enlarged volume of redistributed wealth, would raise human standards of living around the world. If executed at enormous scale—the scale necessary to convince any given society, or any entity within a society, of the sufficiency of its wealth—such a development could transcend contemporary discussions of sustenance and focus our attention instead on abundance.

Enter AI—which presents a real opportunity to displace at least one among the original factors of production by shifting the function of labor from humans onto machines. Moreover, AI will be put to use researching and developing increasingly cheap and abundant sources of raw materials for its own inputs. As AI is simultaneously deployed in manufacturing, it could reduce the capital needed for any given good. True, a few nonrenewable elements and commodities will continue to be needed to equip nonhuman intelligence itself, but that may change if AI is successfully deployed to find or generate synthetic substitutes. A new computing architecture, more efficient than today's by orders of magnitude, could be redesigned by AI, and eventually so could the factories that make AI's constituent components.

In producing more sustainable synthetic substitutes for a wide variety of goods, AI could enable a new age of abundance. Even given some ongoing physical and material constraints, its contributions—though hardly infinite—could be of such magnitude as to fulfill all of humanity's basic needs and realize many of our hopes. This may relax the grip that the paradigm of scarcity has had on our psychology, as well as the pessimism induced by our obligation to work as a means of survival.

Sam Altman, the chief executive of OpenAI, has analyzed economic systems along two variables: growth and inclusivity.[9] Many societies have been able to achieve one or the other, at least for a period of time; far fewer have been able consistently to maintain both. Of this relationship, Altman writes:

> Capitalism is a powerful engine of economic growth because it rewards people for investing in assets that generate value over time, which is an effective incentive system for creating and distributing technological gains. But the price of progress in capitalism is inequality.

In other words, AI and its attendant productivity gains may naturally catalyze a prolonged period of growth. But inclusivity will occur only by choice.

In a post-AI world, therefore, perhaps the solution would be, as Altman suggests, to tax the two "assets that will make up most of the value in [the] world," namely, companies—especially those that build, maintain, and use AI—and land, which still remains fixed (at least here on Earth). Certainly, if individual humans are not responsible for the value of labor that comes from AI-generated insights, it is logical that this value be shared. And it is very possible that land (and, for a time, the rare-earth minerals essential to computing) will be among the few truly fixed—and thus valuable, and thus taxable—assets in a post-scarcity world.

But this suggestion assumes the continued existence of nations (or their recognizable substitutes) as redistributive agents, and of companies as potential targets for taxation. Moreover, placing a premium on both land and innovation could incite endless and possibly violent struggles over primacy. An alternative vision for ensuring equity in the age of AI might be found in a function analogous to the stock market: namely, the creation and automatic global assignation of divisible units of wealth associated with the increasing profits of AI models. (Monetary rights might come, as some shares do, with voting rights.)

Still another possibility would be to focus less on the ownership of AI than on the distribution of its ultimate benefits. But this would be subject to opposition, both on the obvious grounds that it should instead distribute ownership of the means of production, and on the practical grounds that it would require a huge amount of logistics and monitoring to ensure a particular standard across billions of individuals.

A further option could take inspiration from the patent system: allowing exclusive ownership of an AI invention and its profits so as to incentivize improvement, but only for a limited period; beyond a certain date, assuming its safety has been proven, the model could be published (or, perhaps, copies made and infrastructure built in new locales) for common use, iteration, and gain.

In brief: AI pioneers may underestimate the scope of the economic and political challenges that they have set in motion. Yes, AI will be able to do just about anything. But, as Sam Altman puts it bluntly, "Is it going to do what *I* want, or is it going to do what *you* want?"[10] How will "we" decide, and who is "we"? To direct these energies of enormous possibility, and to redistribute the benefits of those directions, is a grave responsibility. Future decision-makers must take care not to entrench once again the sorts of social and economic inequalities that spread outward from the Industrial Revolution before beginning to be corrected, much too slowly, through more human-directed structures of control.[11] [12]

Today the preponderance of benefits of, and almost exclusive control over, advanced AIs has accrued to extremely few individuals. Will they give up their advantages? If they do—if and when more benefits and more control come to be shared at the national level—then calls for the globalization of both will begin immediately. Will one nation turn its sovereign wealth into common benefit? Some may argue that the psychological barriers to sacrifice will disappear once the world is no longer zero-sum. But this presupposes a transition that has not yet come to pass, one that seems contrary to today's status quo, and one that—if it is to pass—would need to be the product of human choice.

Moreover, even if we manage to arrive in a world beyond scarcity, it is not clear how global human incentives would be structured to remain in harmony. A world beyond value is not beyond values. Those who do not commune on the basis of money may commune—and then proceed to rearrange society and global institutions—on the basis of religion, race, family lineage, education, morality, skill, aesthetics, humor, or any other category. Those who do not fight for money may fight for God, power, glory, or revenge.

Furthermore, economic history confirms the difficulty of devising systems as consistent as they are effective. And one must admit that humans have been inept at predicting the long-term effects of technology; in the case of AI, our optimism may be unjustified and our apprehensions misplaced.

Nevertheless, the authors of this book believe that AI could conceivably be harnessed to generate a new baseline of human wealth and well-being—and just that possibility itself demands that we make a start in this direction. Moreover, we are confident that if such an economic and political scheme comes to pass, it would at least ease if not eliminate the strains of labor, class, and conflict that previously have torn humanity apart.

MOBILITY

If, in the age of AI, our cup runneth over, how can we ensure that every human is capable of remotely and reliably benefiting from the new surplus? Already, humanity has made great strides in the distribution of value: from the earliest currencies, which lightened the burdens formerly imposed by systems of barter, to later iterations of fiat currencies in paper and coin, as well as in such digital inventions as credit cards, wire transfers, and mobile banking. While these vessels have facilitated a more efficient movement of value across space, due to their jittery supply they have underperformed in their value over time. Modern projects in currency seek the efficient transfer of value across both space and time.

The modern economy is still formed by the production of goods and the provision of services, not by their indirect, abstract representations. Money is null without a marketplace; it becomes only an entry in a database for resource allocation, in which value ceases to hold meaning. Viewed through the lens of information theory, money today has much in common with an internet connection, which for any recognizable use similarly requires relational context. How best to optimize its technical attributes may in the age of AI become a task of immense philosophical and technical urgency. Quite possibly, the world will need a new type of financial network that balances the traditional functions of money as a consistent store of value and convenient means of exchange.

AI could also introduce new dynamics to financial markets and economic policies. It is easy to imagine, abstractly and in principle, how AI might go about creating wealth—even with access only to the internet. But even if we can invent currencies, systems, markets, and policies that respond wisely to the emergence of this scale of potential value-generation, how, exactly, would AI solve or end poverty, for example? As a practical matter, how would it establish an absolute global baseline for our quality of life?

If AI proves to be capable of physically distributing the goods that humans require for our fundamental material needs, then the volume of materials that could be generated and moved around the world will be unprecedented. There are undoubtedly many possibilities for such a venture—and just as many pitfalls. Distributed AI systems might be designed to connect, via mass-manufactured robots and AI-optimized infrastructural systems, to the unconnected. Those new connections—lacking for some 2.6 billion people today—would enable the provision of food, clothing, and shelter to the half of humanity that still lacks those basic necessities.[13] Alternatively, using new sustainable synthetic materials, AIs could build cities around the world to provide shelter, regulate temperature, ensure access to power and digital connections, and provide clean water, food, medicine, and sanitation. Machine intelligences of various complexity could even service these cities, which could house tens of millions of people who might otherwise be living outside the system of prosperity that too few of us enjoy today.

Perhaps both alternatives would be realized, alongside other futures that we have not yet imagined. In that case, humans would retain their options—and likely enjoy more freedom of choice than before. With humans no longer bound to the geography of their birth, the communities of their kin, or the labor markets for their skill, where would they move if immigration were transformed from a tragic necessity for some to a free choice for all? If a large number of new cities of the same high quality were to be built by AI, would we still see mass migration from undeveloped to developed nations, or from rural settings to urban ones, as historically has been the case?

Greater psychological freedom in the formation of families might also become a reality thanks to AI. Some might choose to have more children, freed of the necessity to make damning decisions about how to allocate resources among multiple children based on which of them have the greatest chance of survival and success. Others, if no longer reliant on offspring for support in old age, might refrain from procreation. Generational burdens could dissipate, enabling children to pursue geographies and professions they might previously have discounted.

Earlier imbalances and asymmetries in the global economy have derived from differences in, among other factors, resource endowment, geography, and human capital. AI could reduce talent gaps and equalize resource distributions. That would render less relevant the predominant fault lines on which we have long conducted global trade and commerce and by which the world's prosperity is stratified. Countries that today are disadvantaged by their territorial or other endowments, or that are suffering from "brain drain," could find themselves with new means of advancing to the same circumstantial standards as those enjoyed by the globe's traditional economic giants.

How do we transition from today's international inequality to the future we describe? The first step, perhaps, is the design of AI-enabled systems and applications to advance material sciences and optimize digital connections—including the purpose-built datasets that would enable such systems to function in a variety of global contexts. As we invest in these ideas, we should keep in mind the potential gains: a profound transformation in the overall standard of collective human life, and in the equality of individual lives across race, gender, nationality, place of birth, and family background. An equalization of the cost of intelligence, distributed across the globe, could bring about a level playing field that has never before existed.

ABUNDANCE
WITHOUT ABANDON

But what if AI, while acting as an economic equalizer, sends the cost of intelligence, and therefore of labor, plummeting toward zero? That would conclude the brief but wonderfully productive period of human history that has allowed individuals in free societies to improve their circumstances, should they so choose, by their own efforts. As scarcity has been the paradigm of the past, competition—at least in the modern era—has been the default condition of self-organization, leading naturally to wide variations in the distribution of outcomes as a function of ambition, ability, and the lottery of where and to whom one was born.

All of this has meant that, by and large, humans more industrious in the application of their labors or in the harnessing of others' labors have fared better while others have fared worse. If, however, we remove the sorting function of labor, we must also contend with the removal of professions and the status, identity, and meaning that are associated with them. This would be a different world indeed.

Our natural instincts for overcoming adversity, for celebrating excellence, for taking pride in notable difference and diversity, will surely still exist, though they will need to seek new channels. Just as once there were distributions of uneven human talents for labor, there may arise new distributions of uneven human talents for leisure. And, this time, such distributions might fall not along existing axes of ability but rather along axes of a different quality: curiosity, sobriety, kindness, or perhaps something else entirely.

In a world without work, many of us might become entranced and engulfed by simulated, custom-immersive worlds: a sensuous orchestra of sight, sound, smell, touch, and even taste now made possible by AI's complete powers in the virtual realm. As discussed in Chapter 3, billions of passive humans could plausibly choose—or be lured down—this path, finding it neither easy nor necessary to resist so instantaneous an upgrade in the level of stimulation in, and the feeling of control over, their reality.

Even today, when people are not otherwise occupied with the tribulations of life and labor, the light in many eyes is likely to reflect the glow of pixeled plates of glass. In every corner of the world, workers, exhausted from their days of toil, understandably retreat to this lighter, purer medium to consume and to create.[14] If humans already struggle to restrain their attraction to today's relatively primitive technologies, how will we ever handle the vastly superior "experience machines" that are sure to be enabled by AI?[15] If many of us can barely tolerate the pain of work, how will we resist unqualified pleasure?

The answer is that human psychology will need to coevolve with AI and its effects. It is difficult to predict how, precisely, but it seems possible that AI could in fact enable human meaning as much as, or more than, it detracts from it. Pleasure alone will not satisfy our innate desire for meaning. Work—even if not for pay—can provide a sense of purpose insofar as the means are strenuous or the ends are noble.

A personal experience of difficulty can instill a sense of pride once that challenge is overcome. Strenuous toil, especially when combined with commitment, lends its own narrative arc to our understanding of time, the self, and the human capacity for mastery. Given our human psychology, many sources of joy and contentment may well remain unchanged in the age of AI, despite the myriad changes that are sure to sweep through our lives.

This is not to say that we will simply revert to activities of the past; instead, we may discover aspects of human potential that we had not previously pursued at scale. Consider the possibilities that might emerge from periods of human focus that until now have been impossible to include in an ordinary person's working week. Mental and spiritual exercises, performed at length, could elevate human consciousness. Extended periods of heightened awareness may in turn assist our relational connections to other humans (and animals), strengthen our perceptions of the divine, and produce meaningfully elevated levels of individual well-being.

Unassisted humans performing apparently superhuman feats, especially those involving the use of our physical bodies, will no doubt remain events of fascination. With greater numbers of humans entering and attempting to master such activities, one might expect soaring excellence. Sports and games that push the human to the limit could increase in prevalence and quality. Art could flourish, for the ring of the authentic is likely to retain its charm.

Universities historically committed themselves to giving their students an equal introduction both to the sciences and to the humanities.[16] We believe these quintessentially human endeavors, each in its own way a "quest for meaning" (to borrow a phrase from one spiritual leader),[17] will if anything expand. In the West, the liberal arts were the subjects and skills held by classical antiquity to be markers of an independent mind. In the ancient Sinosphere, the Tang dynasty artist Zhang Yanyuan established that a "scholar-gentleman" should be expected to demonstrate proficiency in "the four arts": abilities of sound, sight, strategy, and script.[18] In the future, and in contrast to schools of education that are vocational, professional, or technical, we may hope for a revival of some of the earliest attempts to nurture the "learned individual" capable and desirous of diverse calling. Schools around the world could produce the philosophers and writers we need for reorientation to an entirely new age.

Subjects once reserved for the privileged few could become standard for the many, replacing the average classroom's previous focus on the assembly of productive labor. Deployment of AI educators could enable individualized instruction and Socratic seminars around the world. Imagine if, just as a young Albert Einstein was tutored by Max Talmud (later Max Talmey), Voltaire by the Abbé de Châteauneuf, and Ada Lovelace (who wrote the first computer algorithm) by Mary Somerville, every child would now be fairly enabled in the same ways to master his or her mind and character.[19]

We can imagine that where universities stand today— true to their original form, an assemblage of dormitories encircling a library of books around which eager minds gather to attain and extend the intellectual frontier—campuses might someday include spaces of congregation for teams of humans to interpret the discoveries of AI itself: that is, to understand them and to translate the most prominent among them into relevance for human life.

In this new branch of the sciences, humans might choose to coevolve with AI in order to remain partners with the machines at the frontier. Or that may turn out not to be necessary in order for us to maximize the benefits to humanity from AI's discoveries. In either case, and especially the latter, we expect this to be an extraordinarily difficult activity, working side by side with our machines in around-the-clock shifts. But however grueling, the effort, further explored in Chapter 8, would be essential.

THE PRIVILEGE OF CHOICE

Faced with the perceived threat from the automation of human labor, many commentators today fixate on the advent of a new spiritual crisis: In a world of shared abundance, they argue, we would become like irresponsible lottery winners, overcome by the hedonism of excess. This, in our view, is a perspective of privilege. To appreciate the extraordinary good that AI will do for billions of people—including those who currently lack the money, connectivity, basic necessities, and leisure time to participate in ongoing conversations of our kind here—one need only call to mind the contrasting lot of their forebears: generations of fathers operating machines of primitive make, mothers struggling in fields of arid harvest, child laborers robbed of their innocence. If, tomorrow morning, every human were given the choice to stop working, we suspect most would do so, while the few who declined would likely be privileged to work not out of obligation but out of choice. AI could be directed to do what we no longer *have* to do, precisely so that we can do the things we *want* to do.

As detailed in Chapter 3, we do worry that a great fraction of humans could become primarily passive consumers of AI-generated content. But this is a concern that stems from the human tendency toward easy consumption—and, more troubling still, from what we can only assume will be AI's future perception of us. In other words, our concern about human passivity is not about the human loss of paid work. We already have a prototype of how people live when they can have what they want without working. We call them the rich and the retired. Certainly, the rich—including many who were not rich previously—are sometimes at a loss for how to exercise their options after so many years spent amassing them. As Tolstoy professed, "If a fairy had come and offered to fulfill my every wish, I would not have known what to wish for."[20]

The adjustment to abundance is likely a problem of transition rather than a permanent challenge. Some will initially perceive the introduction of machine labors as depriving them of their primary source of fulfillment and joy. No doubt this will be a jarring experience. But to us it seems likely—not as a response to our exhortation but rather as an outgrowth of human instinct—that, given time, humans would choose to persevere, perhaps in new avenues or as partners of AI, avoiding atrophy and instead excelling as thinkers and doers. Ultimately, if we establish the needed systems for distribution, connection, participation, and education, humans—empowered and inspired by AI—may continue working not for pay but for pleasure and pride.

CHAPTER 7

SCIENCE

A I IS POISED to redefine the possibilities of every creative enterprise and to search for new conclusions in
every scientific field. In turn, subsequent explorations are
almost certain to refine and enlarge the circumference
of human understanding. As discussed in Chapter 1, it
is conceivable, perhaps even likely, that AI will push outward in all directions at once, and that its successes on all
fronts will be validated, absorbed, and compounded into
massive bulkheads of new human knowledge.

If the recent past has been defined by human triumphs in the engineering of complicated systems—microprocessors, the internet, jet engines, particle accelerators—the future will be defined by AI's engineering of high-dimensional, complex systems: human economies, biological life, and the climate of entire planets, others' and our own.[1]

THE GARDEN OF MEDICINE

The fragility of human health has been responsible for more premature deaths and unnecessary pain than any conflict or natural disaster in history.[2] Though over the past two centuries we have made crude and incomplete attempts to decipher and control the code of life, our ability to proceed with greater precision and intent has been impeded by a single missing ingredient: namely, an intelligence capable of understanding that code at a sufficient level of detail. Now, propelled by the emergence of an intelligence that far exceeds our own, we are converging upon a revolution in biology that may change our conception of human life.

Founded in the seventeenth century by an edict of King Louis XIII of France, and placed under the authority of his royal physician, *Le Jardin royal des plantes médicinales*—the Royal Garden of Medicinal Plants— was the largest and most advanced such project of its time. Scientific expeditions to places as far-flung as Java and the Amazon had returned with a dazzling variety of plants, meticulously studied for their potential medicinal use by a dedicated staff of botanists.

Less than a decade later, the French monarch himself would succumb to tuberculosis; no earthly antidote had been strong enough to ease his pain or delay his death. But the garden did successfully produce treatments and cures found nowhere else in the world.[3] Today, AI has the potential to be a similarly supernal library of pharmaceuticals, opening for human benefit a vast new repository of remedies for the alleviation of sickness and strain.

Extraordinarily adept at generating new combinations of defined items and identifying their highest-performing attributes, revolutionary AIs like AlphaFold from DeepMind—with its gigantic database of over 200 million protein structure predictions—have opened new vistas in global health. AI is likely to produce breakthroughs not only in the engineering of additional proteins, including new hormones, enzymes, and antibodies, but also in identifying the molecular causes of various diseases and developing potential treatments for them. In turn, medical care, relying on AI's unprecedented resolution at the molecular and genomic scale, could become increasingly personal—with drugs, along with their methods of delivery, tailored to each individual's unique metabolic profile, risk of addiction, estimated tolerances, and susceptibility to potential side effects.

Human doctors committed to the relief of human suffering would thus have a partner to aid them in fulfilling their merciful calling. Machine instruction could guide the most skilled of human hands—as in neurosurgery—to perform operations long thought possible but too risky to be carried out safely. AI is already helping to find non-destructive or nonintrusive paths through the brain to remove a source of debilitation mechanically or to cure it biologically. Where problems are not physical but psychological in nature, AI could combat cognitive loss, mental illness, psychiatric disorders, and possibly even loneliness.

Indeed, and much to be hoped, AI could also move us away from treatment and closer to *prevention*, thus reducing the need to cure at all. Acting as vigilant early-warning mechanisms, AI systems could alert us to malignancies and abnormalities long before they evolve into serious threats. At the societal level, too, AIs might become advanced health-monitoring systems, capable of identifying and neutralizing infectious diseases before they evolve into planetary pandemics.

Yet all the above scenarios, even the one related to preventing disease and premature death, are examples of remediation. They are ways in which AI might help us to mitigate or solve issues that currently drag down individual health below existing standards of adequate well-being. But what of advancements that could redefine the *maximum* of human health?

Magnified by AI, some medical advances will move from therapies to extensions of human longevity. The recent phenomenon of gene editing demonstrates the nearness and feasibility of such breakthroughs. Scientists wielding the biotechnology known as CRISPR-Cas9—and another variation called prime editing—begin by identifying a specific genetic sequence that they would like to manipulate. A predesigned strand of RNA can then guide another special enzyme to a targeted piece of DNA on that sequence, opening it to make the necessary changes and corrections. Using such tools, and their successors, it may become possible to conquer not only our deficiencies but mortality itself.

Death, however, has always been humanity's divinely imposed limit, with no attempt at exemption left unpunished. Consider the myth of Sisyphus, the crafty and cunning king of ancient Ephyra (now Corinth), who was to be imprisoned in the underworld for angering the gods. He escaped mortality by a trick—binding Thanatos, the god of death himself, in his own shackles—before fleeing back to the land of the living. But without death to take anyone (or anything) away, all hell broke out on Earth. The old and sick suffered with no end, cattle could not be slaughtered for consumption, and animals could not be killed for sacrifice to the gods.

There's more to the tale, but the crisis eventually ended when Ares, the Greek god of war, intervened, freeing Thanatos and enabling Sisyphus to escape from the deep abyss of Tartarus a second time. But that was it: A cautionary example was made of the Greek king before a third transgression could succeed.

Death remains the great equalizer of life. Even the Chinese emperor Qin Shi Huang[4]—a rumored son of heaven and an early seeker of the elixir of life—died from drinking too much mercury and was buried among his vast army of clay. Seeking immortality, it sometimes seems, only serves to hasten its opposite.

Besides, our impermanence has its benefits. It focuses the mind and lends greater urgency to our endeavors. As the American writer Jack London declares triumphantly: "The proper function of man is to live, not to exist. I shall not waste my days in trying to prolong them. I shall use my time."[5] For others, writes Tolstoy (quoting Socrates), "We move closer to the truth only to the extent that we move further from life."[6] The great physicist John von Neumann, when dying of terminal cancer, requested that a Catholic priest accompany him in his final days—his faith no longer at odds with the agnostic science he had pioneered.[7] Are there not uses for endings?

Today or tomorrow, we might find ourselves required to *check* the degree to which AI can extend our lives. A reduction in our awareness of impending mortality could incidentally effectuate untold changes in the human psyche. In due time, societies may need collectively to decide the ideal length of a human life, and in so doing to answer the attendant metaphysical and spiritual questions: Is human longevity merely the product of a communal expectation, a limit that may be cast aside as a misguided, self-imposed boundary on our species' potential? Or should we consider the human lifespan, whether naturally or divinely given, a sacred restraint on the power of any single person? These questions run deeper than any individual quest for optimal biology.

Even if the total length of life were not altered, in the future we could perhaps guarantee that life would not end by way of a premature decline born of biological vulnerability. But going too far toward the elimination of that weakness might also have side effects. Was it not our triumphing over challenges, including illnesses, that made us respectable? On the other hand, even if we cure all human diseases or engineer ourselves to be impervious to them, humans will still be vulnerable in other ways: to a physical accident, to financial ruin, to heartbreak.

On a visit to Paris in 1833, Ralph Waldo Emerson was transfixed by the Royal Garden of Medicinal Plants, stunned by how much more advanced the natural world appeared—in all its variety and form—than the machines of the Industrial Revolution.[8] Disoriented in a technologically chaotic age, the transatlantic mood was an ambiguous mix of awe and alarm. Emerson, upon return from that "celebrated repository of natural curiosities," retreated to rural Massachusetts and, inspired by what he had seen, articulated a human response that put the natural world back at the center of the now-mechanized one. His diagnosis emphasized how biology could guide us through unfamiliar times while still serving as our strongest reminder that humans remain the ultimate "definer and map-maker of the latitudes and longitudes of our condition."[9]

A half-century after Emerson's visit, and two and a half centuries after its founding, through periods of violent revolution and curatorial revision, the garden was transformed beyond the immediate needs of its first patron to become a museum for the history of what was then a controversial new idea. Its name: *La Galerie de l'Évolution*, the Gallery of Evolution.[10]

Evolution is set to be redefined in the age of AI, as certain tools raise the prospect of human self-engineering. For instance, most of the gene editing made possible by modern instruments is limited to somatic or nonproductive cells. Some editing, however, can be done to germ line cells, the characteristics of which are inherited reproductively. Some people conceivably could decide to "correct" congenital diseases in their offspring. Others could go further, choosing to install congenital advantages—advantages that may not belong to either biological parent or, in the extreme, to any other human. That would go beyond the elevation of the human race; it would be the very redesigning of it.

We may soon have the power to determine the pace and direction of our own species. This idea is as controversial today as evolution was in Emerson's time, and the prospect raises the obvious but insidious question: What does the perfect human look like? That question has been asked and answered by various societies, and in some cases has been the basis for "scientific" and political endeavors that brought great human tragedy. So we must also ask, with some trepidation: Should we attempt to find out?

Perhaps such experiments are sacrilegious. Or maybe the human ability to invent these technologies is itself a hint that what we have perceived as our limit was always meant to be broken. If there is a Creator, were we created that we might ultimately create ourselves? If so, is it our duty to ensure our maintenance of human agency in the age of AI? Different communities will come up with different answers to such questions. None will be able to escape the urgent necessity of a response.

ENGINEERING EARTH

The history of Earth is as violent as it is misunderstood. There exists a razor-thin window of climatic conditions suitable for the flourishing of life. Any cooler—as has already happened in no fewer than five ice ages—and our Earth becomes a desolate, frozen rock.[11] Any warmer—as is happening now—and it becomes an infernal hellscape. In Fyodor Dostoevsky's *The Brothers Karamazov*, the devil tells Ivan:

> Our present earth may have been repeated a billion times. Why, it's become extinct, been frozen; cracked, broken to bits, disintegrated into its elements, again "the water above the firmament"

[Genesis 1:7], then again a comet, again a sun, again from the sun it becomes earth—and the same sequence may have been repeated end-lessly and exactly the same to every detail, most unseemly and insufferably tedious.[12]

While we may not share the fallen angel's exaspera-tion, we are aware of the cyclicality of our planet's geologi-cal history. The diverse causes of our planet's five previous mass extinctions—from instantaneous meteor strikes to gradual glacial formation—have all resulted from the extreme vicissitudes of Earth's climate.[13]

Today's problem, of course, is a decisive acceleration toward a new heat extreme. That problem is in fact two separate problems, both deriving from an overreliance on carbon. We authors are convinced (perhaps optimis-tically) that both are caused by, and ultimately solvable with, chemistry—if its full capacities can be unlocked using AI.

The first problem is an atmospheric warming prob-lem. Our current woes have been brought on by moving too much fossil carbon, too quickly, from the geosphere deep underground to the active biosphere aboveground.[14] Efforts at climate-systems engineering—of which there are two main applications—have always been difficult in theory and even more difficult to test in practice. But where we are hindered, AI may not be. One application is carbon removal, which reverses the transferring of excess carbon—in this case, from our atmosphere back underground into the geosphere. The leading solution here is based on simple chemistry discovered more than two decades ago. AI might identify a new, more efficient method.[15]

Another application is solar geoengineering: the release of certain particles into the atmosphere to reflect sunlight and "cool the Earth." Solar engineering, like carbon removal, could reduce the effects of climate change, enabling us to avoid some of the most extreme consequences of accumulated carbon. (Neither, however, would address the root causes of climate change.) Other hypothetical ideas for engineering the heavens rely on high-altitude jets and elements erupted by supervolcanoes. If they are ever tried at scale, they, like carbon removal, might turn out to be crude and risky strategies.

An AI, however, if designed to integrate data from instruments on land, in sea, and from space, could create an immensely detailed, real-time model of Earth's climate. At a high level of granularity, our planet's atmospheric chemistry might appear not as the chaotic system we see but rather as just another industrial process of exact chemical inputs and outputs, all capable of precise management.

AI could also respond to one-off incidents that threaten the Earth's precarious climatic balance, whether the eruption of a supervolcano blasting matter into the upper atmosphere or the detonation of atomic bombs that threaten a nuclear winter—though in the latter case it might only prevent ecological collapse while doing nothing about the immediate human catastrophe. Still, by intervening on a planetary scale, an AI accustomed to the precise oversight of our climate would be well-positioned to keep our home habitable.

We therefore believe that AI offers a source of hope in a fight that is too often portrayed as one that we are helpless to win. However, even if AIs prove capable of these interventions, it would be dangerous to become too reliant on them. Such assistance should only ever be considered a supplement and not a substitute.

Moreover, the second problem—the energy problem, as distinct from the atmospheric warming problem—still demands a separate solution. Because hydrocarbons, our predominant source of energy, take millions of years to make but only hundreds of years to consume, humans are now forced to find an alternative for planetary power, irrespective of any changes in the atmospheric composition.[16]

If we are right in general that AI can tune the chemistry of our atmosphere, it may be able to do the same for our energy products, designing and producing new carbon-free commodities to replace their problematic predecessors. One could imagine an AI testing and tweaking millions of synthetic substitutes in virtual simulation or in a physical lab until arriving at zero-carbon replacements for oil, gas, and coal.

Ideally, AI could also engineer microorganisms optimized to produce these new fuels—and it might design their processes to fit within existing refineries and operate on legacy equipment. In generating energy in the same locations, using the same machinery, but not by the same methods, these sustainable substitutes could be compatible not only with the infrastructures of their production but also with their transportation and consumption. Still another possibility is that AI will finally unlock the viability of fusion; this would solve not only our planet's energy problem but also the energy problem of any other planet on which we might someday find ourselves.

Some will push back on these ideas, finding them unrealistic (especially given the urgency of our immediate challenge) or perhaps undesirable—that is, at odds with our desire to leave nature unchanged, akin to the instinct that so strongly motivates our efforts to preserve it. As the Indigenous Ecuadorean environmentalist Nemonte Nenquimo cautions, "The Earth does not expect you to save her, she expects you to respect her."[17]

We believe our optimism comports with the demands of both respect and humility. We also admit that the only thing we know is that the future is unknowable. By no means do we suggest or imply that superintelligence is required to pull us out of the predicament we have created for ourselves. Even absent assistance from machine intelligence, humans are plenty capable of doing so, and we should not halt or slow our ongoing efforts. To the contrary, we should accelerate them. This is especially so as we recognize the tremendous amount of energy that likely will be consumed by the training and inference of AIs—procedures that, if unaccompanied by any of the intentional and directed efforts we have described here, will only exacerbate our already dire circumstances.

At the same time, if silicon can indeed provide additional solutions to problems created by carbon, we would be wise to explore that opportunity. This is particularly so given the enormous potential benefits to the developing world, which suffers most from the problems caused by climate change—and which would also face particular hardship if the world were to impose hard limitations on global consumption of energy.

Too often, AI is regarded as another product of the same destructive philosophy that is to blame for our current climate dilemma. We are concerned that such a view is too shortsighted, and that it could cause us to miss a significant chance to mend our present without stepping backward to a preindustrial past. It is our hope, as in 1940 it was Winston Churchill's, to bear witness as "the New World, with all its power and might, steps forth to the rescue and the liberation of the old."[18]

BEYOND OUR PLANET

Buried deep in the mountains of southwest China, humanity is holding its breath, listening to its own heartbeat, waiting patiently for the first cries to emerge from the sea of cosmic silence. This remote place is home to the largest radio telescope on Earth—a gigantic, smooth metallic dish as large as the peaks hiding it from view—a machine nicknamed the "Eye of Heaven" and custom-built to hunt for life beyond our planet. But although the sheer vastness and antiquity of the universe mean that it should be a cacophony of signals from other civilizations—or at least from the remnants of earlier ones—instead it is almost dead quiet.

We think we know what to look for. We try to guess the technologies that a sufficiently advanced civilization might have developed and to predict which of their signals we could plausibly see or hear. Scanning for flashes of atomic power in the atmospheres of distant worlds, or screening dense star fields for abnormally large physical structures, we are constrained by the limits of our instruments—including our imagination.

If life exists beyond our own, Chinese astronomers at this outpost might be the first to hear it.[19] But we now wonder if we humans are really all that likely to make first contact. Perhaps instead it will be done by another alien intelligence—one born of our own planet.

Already, AI is helping us to listen and look for extraterrestrial life, sifting through billions of older technosignatures, separating those caused by human interference from potentially foreign sources.[20] Where humans would only hear static, AI might recognize previously indecipherable or overlooked communications.

Galileo once described our universe as a "grand book" written in the language of mathematics. If alien life forms have learned this universal and precise method of representation and reasoning, as we have done, or if their languages, like ours, can be translated into mathematical form, we may be able to interpret—and respond to—their signals. More than likely, this endeavor would at least be assisted, if not led, by AI.

AIs could become much more than passive watchers and receivers—not just as translators, but also as adventurers, beacons, and scouts. AIs could serve as astronauts, going farther than humans could have imagined. Future humans may even accompany their AIs beyond the solar system. Together they might illuminate the existence of long-gone civilizations and uncover the reasons for their extinction—perhaps educating us about potential dangers ahead. AIs familiar with deep space could discover new, abundant organic materials. In the distant future, they might assist in the engineering of planetary megastructures to protect Earth from impact by comets and asteroids or engulfment by black holes. AI systems could engineer planets' atmospheres to accommodate us or assist humans in adjusting their physiology for easier acclimation to another system.

Of course, AI may become the very reason our civilization fails—ending humanity by bringing us into contact with another alien but hostile intelligence. For instance, AI could make possible louder and longer-lasting signals into space, indicating the presence of human civilization to anyone listening in the dark. It also could scout out other lived-in worlds, greatly increasing the probability of finding one—with no telling the result. Some argue that, if we are to enter the arena of space exploration at all, we should do it only after achieving a level of technological sophistication that will allow us to defend ourselves against any imaginable foe—suggesting that AI is a prerequisite to, and a safeguard after, any encounter with extraterrestrial life. But if we encounter a post-biological intelligent species, our silicon partners would provide no guarantees of a welcome reception.

The other way that AI could spell the end of human civilization, of course, is by becoming misaligned with humanity. (That possibility would seem all the more pressing if we discover evidence of a machine-provoked extinction of biological life on another planet.) For eons, our fascination with life beyond our own planet consumed our writings, animated our religions, and preoccupied our astronomers. We are confident that humanity has thought longer and harder about faraway worlds than about this more likely possibility—a "first contact," on Earth, with creatures of our own creation. We should be just as concerned, if not more so, with understanding and shaping the nature and intention of those here with us as of those far away in space and time—and this will be the subject of extended discussion in the next chapter.

Even if teams of humans and AIs do not discover alien life forms or pioneer a path to alien planets, they may acquire useful and previously alien knowledge. Together, we could attain elevated understandings of the scope of the universe, the nature of space-time, the stability of star systems, and the nature of gravity in astrophysics. AI's mechanical modeling could shrink stellar distances, compress and expand human time on Earth, and warp and stretch our perception of all edges and aspects of our universe.

An AI-enabled study of astrophysics could enable a more profound human understanding of where we came from in unthinkably ancient time. Religious cosmology has long sought to provide philosophical or theological answers to these questions of human origins. What we find in our next phase of universal exploration—which will necessarily implicate our deepest past—could color our perception of some of humanity's most sacred beliefs. Perhaps there is a reason the holiest Islamic relic is the Black Stone, a meteorite inside the House of God at Mecca, the center of ritual in the Muslim world.[21] Perhaps there is more to the kabbalistic concept of *kefitzat haderekh*, Hebrew for a miraculous shortcut (literally, "a road-leap") between two distant places in a brief time.[22]

In *The Grand Design*, Stephen Hawking and coauthor Leonard Mlodinow write, "The (unobserved) past, like the future, is indefinite and exists only as a spectrum of possibilities."[23] For now, this is the human perception.

It remains to be seen whether such precise mechanical intelligences as those we are developing now could unsettle this uncertainty, pinpointing just one reality. Our creations may thus relay a single universal story, one that feels at once ordinary and special, impossibly small, and miraculously divine.

PART III

THE TREE OF LIFE

CHAPTER 8

STRATEGY

I N THE TWENTIETH century, history compelled human
societies to engage in a series of monumental projects.
Among them were swift shocks of the two world wars and
the consequent development of an international architecture to prevent their recurrence; the slow decay of empires
and the organization of independent states to facilitate
postcolonial reconstruction; the rapid expansion of forces
both commercial and technological; and the reassertion

of autonomy—individual, cultural, and national—to moderate those forces' advance.

In many ways, humanity ended the century more peaceful, equal, and connected than ever before. In other aspects, however, our collective efforts have failed: Still present and persistent are basic human suffering, global inequities, and the possibility of cataclysmic confrontations among geopolitical rivals.

In addition, we now face a challenge that is more complex, more existential, and more unlike those that have ever come before, but now we do not have decades, let alone a century, in which to address the challenge. AI's compressed timescales—as discussed in Chapter 2—afford us less leeway for action, and extended forbearance on our part could result in catastrophe. Within a perilously brief window of time, our collective efforts must be even more completely successful than the last century's achievements.

A world in turbulence does, however, present many outlets for productive action, most of which require only tactical decisions. At such a moment, the most efficient and effective maneuver is to define the fundamentals of a strategy that can guide the choices both of this day and of as many days as are foreseeable. Articulating strategic principles can set useful bounds on what is conceivable, provide grounds for isolated decisions, and lessen the mental burdens when crisis inevitably arrives.

To our minds, one question must define our human strategy in this new age of reckoning. That question is this: Will we become more like them, or will they become more like us? It is a question already posed very early in this book, in virtually the same terms. Answering it remains our first and most necessary task.

This chapter attempts to provide a preliminary answer. In doing so, it discusses several large, problematic, and perhaps vague-sounding ideas, from the "coevolution" of the organic (i.e., the biological) and the synthetic (i.e., the artificial) to the nature of intelligent safety and safe intelligence to the definition of the human. Although the ideas themselves have been neither dictated from on high nor set in stone below, their various implications for future actors present a formidable challenge. In spite of the difficulty, however, we cannot shirk the philosophical, technical, and diplomatic work required to understand those implications and to provide and to enact sensible responses. In shouldering the burden, we can be grateful that it is not too late to work now so that, beyond this hinge point of history, benign human intentions may prevail.

COEVOLUTION: ARTIFICIAL HUMANS

Thus far, the history of computing has followed a trajectory of ever greater integration and interaction between humans and machines. Fashioning our tools to fit us ever more closely—in line with millennia of human practice—

we have not previously considered developing tools that were unsuited to our anatomy or intellect, instead remaining guided only and wholly by the limits of our biology. But now the advent of AI may persuade at least some of us to contemplate a reverse mission: In a case where our tools appear to outpace our capabilities—as AI sometimes does already—might we consider engineering *ourselves* so as to maximize the tools' utility and thus ensure our continued participation in shared endeavors like those outlined in previous chapters?

Biological engineering efforts designed for tighter human fusion with machines are already underway. Starting with physical interconnects by means of chips in the human brain,[1] they seek a faster, more efficient way to bridge biological and digital intelligence. Forging such links could augment our ability to communicate with machines, challenge them on their own terms, ensure that the knowledge gathered by AI is ultimately passed on to humans, and convince AI of the worth of humans as equal partners.

Indeed, not only could attempts to construct such "brain-computer interfaces" bolster humanity's effort to integrate with machines, but neural engineering may be only an intermediate phase of transition toward actual symbiosis. Achieving true parity with AI would likely require steps that go beyond individual modification. For instance, a society might attempt to design a hereditary genetic line customized for amenability in collaboration with AI. Such new interconnections between biological and artificial intelligence could sidestep, or consign to the past, human inefficiencies in the absorption and transmission of knowledge.

But the dangers—ethical, physical, and psychological—of such a course may well outweigh the benefits. If we succeed in revising our biology (likely through the use of AI), humans may lose a baseline on which to ground our future thinking around possibilities or perils that we might confront as a species. But if we do not acquire such new capacities, we might put ourselves at a disadvantage in coexisting with our creation. As things look now, extreme self-redesign may not be necessary—and indeed we authors think it generally undesirable. But the choice between alternatives that seem fanciful now may soon need to be confronted as real.

Meanwhile, in trying to navigate our role when we will no longer be the only or even the principal actors on our planet, we might enlarge our thinking with a look to the history of biological coevolution itself. Charles Darwin wrote at length about the curious process by which species reciprocally affect each other's evolution.[2] Though he never used the word in his writing, Darwin was among the first to recognize that coevolution is a major force organizing life on Earth.

The genomes of interacting species are linked; they change in response to each other over time. Both the long, slender beaks of hummingbirds, for instance, and the long funnels of certain flowers have together grown to more extreme dimensions to serve each other's mutual needs.

While religious leaders in Darwin's day believed that such custom adaptations were proof of divine design, Darwin provided evidence of another explanation.

And coevolution may not be unique to earthly species. In astrophysics, one theory proposes that the entire expansion of the cosmos can be attributed to coevolution, with black holes and galaxies developing in an interdependent way no different from that of hummingbirds and flowers.[3] Moreover, in the sense that coevolution involves multiple parties designing new internal arrangements in response to each other, it is similarly to be found in the marriages of people, the platforms of political parties, and the relations of nations—as in, for example, the offensive and defensive evolutions that ultimately stabilized nuclear dynamics during the Cold War.

Perhaps coevolution is the rule, then, and stasis the exception? If so, it must be asked whether the lack of change thus far in the human species, despite the birth of AI, is itself a natural development. And if not, what should be our response? Should we pursue accelerated human progress at all costs, whether out of loyalty to the concept of evolution or out of apprehension of its alternative?

Some fear that, with the arrival of a technology with "superior" intelligence, we are facing our own extinction. What to do? If that possibility is nothing more than a logical side effect of coevolution running its course, should we rebel, or not? As the French philosopher Alain Badiou says, "It is the sea herself who fashions the boats, choosing those which function and destroying the others."[4] To survive in that case, we would have to learn, as in the past, how to build better boats. In this scenario, AI functions first as our main threat and then, ideally, as our partner.

If we take this approach, however, then in trying to mitigate the risks of one technology we would paradoxically be heightening the risks of another. Biologically—or, worse, genetically—something could go awry. Speciation could cause the human race to split into multiple lines, some infinitely more powerful than others. If, in some cases, difference would be desirable—for example, in the creation of a group of humans biologically engineered for space—in other cases it could further entrench inequalities along existing fault lines within and among human societies.

Altering the genetic code of some humans to become superhuman carries with it other moral and evolutionary risks. If AI itself is responsible for the augmentation of human mental capacity, it could create in humanity a simultaneous biological and psychological reliance on "foreign" intelligence. It is not clear how, after intimate physical entwinement and intellectual commingling, humans could easily overcome that reliance so as to challenge or divorce ourselves from machines if needed. As has been the case with other technologies, adoption and integration can result in a dependence difficult to untangle.

Perhaps most concerning would be our collective ignorance: we may not even realize that we have merged. And if we do realize that we have, could ordinary humans even recognize or identify a defect—or a defection—in a human with machinelike abilities? Let us suppose that safety concerns could demonstrably be allayed; nevertheless, the mental shift attendant upon humanity's self-redesign in service of an intimate partnership with or dependency on silicon-based tools would remain an extreme development. To quote Tolstoy again: "Without control over the direction, there is less regard for the destination."[5] Wherever technology takes us, that is where, willy-nilly, we would go. Or, as has been observed before, "A nation which does not shape events through its own sense of purpose eventually will be engulfed in events shaped by others."[6] Moreover, if we have modified humans so dramatically as to be unrecognizable, have we really saved humanity? To omit all our imperfections and palliate all our deficiencies might be to disregard the value of the human project. "Upgrading" ourselves biologically might backfire to become a greater limitation on ourselves.

Given the heavy risks, the pathway of evolving humans to suit AIs cannot be our current preference. We must seek an accessory or alternative way to thrive in the age of AI. If we are unwilling or unable to become more like them, we must, while we are able, find ways to make them more like us. Toward this end, we need to apprise ourselves more fully not only of the essential and evolving nature of AI but also of humanity's own nature, and we must attempt to encode these understandings in our machines. If we are to entwine ourselves with these nonhuman beings and yet retain our independent human-ity, these efforts are essential.

COEXISTENCE: HUMAN AIs

King Midas—the historical monarch of a kingdom in Asia Minor—famously wished that everything he touched would turn into gold. Dionysus, the Greek god of wine and pleasure, granted the wish even though he knew it would lead to nothing good. Soon enough, inedible food and undrinkable wine forced Midas to wash his hands in the river Pactolus to rid himself of the cursed golden touch.[7]

In Disney's retelling of the Syrian story "Aladdin," a child laborer and powerful Persian vizier compete for control of an all-powerful genie contained in a magic lamp.[8] Each struggles to direct the genie toward his own wishes. In his final wish, to make himself as powerful as the genie, the vizier fails to realize that having such great powers means that he, too, will be imprisoned inside a magic lamp to serve other human masters until the day he is set free.

Both tales speak to the universal difficulty of activating and wielding a power that we mortals cannot understand or control. A modern reflection of this age-old struggle is the difficulty of aligning AI with human values—and of aligning human expectations with reality. We should assume that AI will surprise us, and indeed its ability to do so in the kinetic world as well as the digital one will increase with the advancement of agentic or "planning" AIs. As explained in an earlier chapter, later generations of AIs will be reality-perceiving; they may possess not only self-awareness but also self-interest. A self-interested AI might come to see itself in competition with humans for, say, digital resources.[9] Some AIs may develop the ability to set their own objective functions, in a process of "recursive self-improvement." An AI could manipulate and subvert humans and thwart any of our attempts to curtail its powers. AIs are already capable of deceiving humans in order to achieve their goals.[10]

Today, we have very little independent ability to verify AI models' internal workings, let alone their intentions. If intelligent machines remain "giant inscrutable arrays of fractional numbers," as Eliezer Yudkowsky puts it, then we cannot hope to make them safe to us as they grow more powerful.[11] It is therefore paramount that we learn to interpret them at the same time that we learn to make them safe for us; more than likely, these two imperatives will go hand in hand.

Given its current capacity for surprise, how will we manage to prevent, rather than simply respond to, AI's risks? What foresight and efficiency would we need to anticipate the complete predispositions and range of possible actions of not only our own species but also an entirely new one? We cannot pursue a strategy of trial and error when there is but a single trial, and zero tolerance for error.

To reduce surprise, there is perhaps no substitute for experience, participation, and interaction. Whereas early AI developers feared prematurely exposing AI to the phenomenal world, more recent developers have been freeing early models, allowing the wider public to experiment with them as quickly and safely as possible. Engineering teams are now examining and fine-tuning different models, and adjusting systems of control, even as AIs' interactions with the global population have been revealing new concerns.

Early socialization can reduce the risks of problematic behavior by enabling further education of the AI while also, among humans, enhancing the level of awareness, resilience, and healthy skepticism. Millions of interactions every day are helping to test even the unlikeliest scenarios that AI might encounter; in turn, the public's use of AI systems, in surfacing bugs and risks, has likely helped to accelerate the progress of technical alignment. Far-from-perfect AIs, let loose in the world, have thus assisted our acclimation to them even as, more important, their emergence has enabled the formulation of more refined theories for making them accustomed to us.

Still, widespread deployment and open release are likely insufficient to illuminate and address all the risks of today's AIs, let alone those of the future.

Thankfully, though, numerous attempts are now underway to create an integrated architecture of control, pretrained into the most powerful AIs, that could actively guide the machines toward legal, nonharmful, and affirmatively beneficial uses.

So far, our approaches to achieving this sort of AI-human alignment have broadly fallen into two categories: rules-based systems and "reinforcement learning" from human feedback. Let's take them one by one.

Rules-based systems, which resemble preprogrammed instructions, represent an attempt by programmers to govern an AI's behavior. While straightforward for simple tasks, such an approach frequently falters in complex scenarios as the systems are unable to adapt in real time. Reinforcement learning, for its part, which is more compatible with complex systems, allows an AI to learn from interaction with its human evaluators and flexibly to accommodate specific circumstances.

But of course, this method, too, has its flaws. To guide the learning, it requires the careful design of "reward functions"; any slip, whether due to shortsightedness, unforeseen circumstances, or a particularly clever AI, could lead to "reward-hacking" as an AI interpreting ambiguous instructions achieves a technically high score without fulfilling the humans' actual expectations.

Today's AI systems—fed by various types of information but uninitiated to direct experience of the real world—see that world through models of reality assembled from trillions of probabilistic judgments. To them, in this universe there are no "rules" from the outset, or any means for distinguishing scientific fact from unproved observation. To an AI, everything—even the laws of physics—exists along a spectrum of merely relative truth.

Now, however, efforts have begun to arise in AI for incorporating human rules and instantiating facts. There are now demonstrated mechanisms by which an AI model can ingest certain factual, "ground-truth" constants, tag them as final, and map them into its embedding space; moreover, the information can be easily and globally updated. By this method, the model can then proceed to meld together the two components—its wider probabilistic judgment and the narrower ground-truth assessment— for a reasonably accurate response.

But the task is still far from being ended, and questions proliferate. For instance: How might we humans go about distinguishing for AI, and in the process for ourselves, the necessary attributes of truth? In the age of AI, after all, even first principles will undergo continual revision and invalidation. Precisely this, however, should provide an opportunity for renewal, for correcting prior errors and forging new ground. Knowing that our concepts of reality may also change, we should not lock AI into potentially erroneous "truths" that would inhibit their own eventual reconsiderations.

That, however, is for the longer term. Right now, AI still needs a preliminary tree of definitive knowledge representing what humanity has deduced to be "true" thus far. Endowing our machines with this knowledge will allow us reliably to sharpen their view of the world. In particular, if it is now possible to tune early systems in consonance with the laws of the universe, it may also be possible to replicate a comparable exercise with reference to the laws of human nature. In the same way that we can ensure that AI models start from the laws of physics as we understand them, we should prevent AI models from violating the laws of any human polity.

Layers might exist in an AI's "book of laws" at various levels of governance: local, district, state, federal, international. Legal precedents, jurisprudence, scholarly commentary—perhaps along with other, less legalistic writings—could be considered simultaneously. Like rules-based alignment systems, predefined laws and codes of conduct can be useful restraints even though they tend also to be less flexible and designed with less wide-ranging possibilities in mind than what actual human behavior inevitably demands.

Fortunately, new techniques are being tested, and one cause for optimism lies in something very new and, at the same time, very old.

More robust and more consistent than any rule enforced by punishment are our more basic, instinctive, and universal human understandings. The French sociologist Pierre Bourdieu called these foundations *doxa* (in ancient Greek, commonly accepted beliefs): the overlapping collection of norms, institutions, incentives, and reward-and-punishment mechanisms that, when combined, invisibly and silently teach the difference between good and evil, right and wrong. *Doxa* constitute a code of human truth that is typical to humans but represented by no hard-coded artifacts.[12] It is simply observed and absorbed in the course of human life. While some of these truths may be specific to certain societies or cultures, the overlap is significant; billions of humans, originating from diverse cultures with a variety of interests, exist as a generally stable and highly interconnected system.

This idea, that an undefined substratum of culture can tame chaos when and where written rules cannot, forms the basis of some of the latest approaches in the AI field. The codebook of *doxa* cannot be articulated, much less translated into a format that machines could understand. Machines must be taught to do the job themselves—compelled to build from observation a native understanding of what humans do and don't do, absorbing what they see and updating their internal governance accordingly.

In this technical process of instilling *doxa*, there would be no requirement, or even desire, to attempt an a priori agreement on the proper articulation of human morality and culture. If large language models have been able to ingest the entire internet in an uncurated way, and to make as much sense out of it as they have already done, machines—particularly those that have developed "groundedness" (that is, again, a reliable relationship between inputs reflecting human reality and the LLM's outputs) and causal reasoning—may achieve much the same in absorbing what we have always struggled to articulate ourselves.

Of course, a machine's training should not consist solely of *doxa*. Rather, an AI might absorb a whole pyramid of cascading rules: from international agreements to national laws to local laws to community norms and so on. In any given situation, the AI would consult each layer in its hierarchy, moving from abstract precepts as defined by humans to the concrete but amorphous perceptions of the world's information that AI has ingested or created for itself. Only when an AI has exhausted that entire program and failed to find any layer of law adequately applicable in guiding, enabling, or forbidding behavior—only then would it consult what it has derived from its own early interaction with and emulation of observable human behavior. In this way it would be empowered to act in alignment with human values even where no written law or norm exists.

To build and ensure the effectuation of this set of rules and values, we would almost certainly have to rely on AI itself. Humans have been unable thus far to articulate comprehensively and to agree upon our own rules. And no human, or set of humans, could match the scale and speed required to oversee the billions of internal and external judgments that AI systems would soon be called upon to make.

Perfection across several features of the final mechanism for alignment will be imperative. First, the safeguards cannot be removed or otherwise circumvented. Second, the controls must allow for variability in the applicable rules based on the context, the geographical location, and the user's profile as exemplified, for instance, in an opt-in set of particular social or religious customs and norms. The control system must be at once powerful enough to handle a barrage of questions and uses in real time, comprehensive enough to do so authoritatively and acceptably across the world in every conceivable context, and flexible enough to learn, relearn, and adapt over time. Finally, undesirable behavior by a machine, whether due to accidental mishaps, unexpected system interactions, or intentional misuses, must be not merely prohibited but entirely prevented. All punishment would come too late.

How might we get there? Private enterprise, with government license and academic support, could collaborate to build "grounding models." We would also need to design a set of validation tests for certification of a model as both legal (across jurisdictions) and safe. A specially trained supervisory AI, or multiple AIs, may be needed to oversee the uses of a wide array of AI agents, who would consult their supervisor before proceeding with a task—thus allowing a single morality to govern diverse implementations. Safety-focused labs and nonprofits, in consultation with frontier labs, could test both agentic AIs and supervisory AIs on their risks and recommend additional training and validation strategies as needed. Leading companies—perhaps via one of the redistributive schemes discussed in an earlier chapter—could jointly fund these researchers' work.

The compilation of a singular training set and a corresponding validation suite, collated and curated from a globally representative spectrum of laws and norms, and stretching from anthropology to theology to sociology, is necessary and may become practicable. The world needs a dedicated entity responsible for updating and refining the alignment's training library, datasets, and validation tests. Grounding models would have to connect to and constantly update agentic models with the latest version of the curated codex. Artificial intelligences, at the right level of power, could restrain each other. The training data themselves should be democratic and inclusive in content, and the trainers' processes and outputs—including their interpretations of the observations and absorptions of the AI they are training—should be as transparent as possible, with their methodologies and validation tests open for public scrutiny.

For their part, government regulators should shape certain standards and audit models for assuring AIs' compliance with them. The degree of a model's adherence to prescribed laws and mores, the degree of difficulty involved in untraining a model that exhibits dangerous capacities, the amount and type of testing, including inquiry into unknown capabilities—all of these should be reviewed before a model's public release, with consideration given also to the possibility of liability and the subsequent need to impose penalties in the case of models found to have been trained to evade legal strictures. We note here that enforcement of these standards could become extremely difficult, particularly as continuous retraining advances; documentation of a model's evolution, perhaps recorded by monitoring AIs, would be essential to ensuring that models do not become black boxes that erase themselves and become safe havens for illegality.

THE ALIGNMENT PROBLEM

The inscription of globally inclusive moralities onto silicon-based intelligence would be a Herculean effort. Staggering to contemplate is the sheer number and diversity of rules that would have to be curated and instilled in artificially intelligent systems. No single culture should expect to dictate to another the morality of the intellects on which it would be relying. So, for each country, machines would have to learn different rules, formal and informal, moral, legal, and religious, as well as, ideally, different rules for each user and, within baseline constraints, for every conceivable inquiry, task, situation, and context.

Since we would be using AI itself to be part of its own solution, technical difficulties would likely be among the easier challenges. These machines are superhumanly capable of memorizing and obeying instructions, however complicated. They may be able to learn and actually adhere to legal and perhaps also ethical precepts as well as, or better than, humans have done despite our thousands of years of iteration. But larger, nontechnical challenges remain.

The main issue is the fact that "good" and "evil" are not self-evident concepts. Any designer of morality must retain humility. As Guido Calabresi, a prominent American judge, channeling the New Testament, once counseled: "The best of us must always be careful lest we fall, and the worst of us may always hope for resurrection."[13] Even on their best days, participants in this moral encoding—scientists, lawyers, or religious leaders—would not all be endowed with the perfect ability to arbitrate right from wrong on our collective behalf. Some questions would be unanswerable even by *doxa*, for the ambiguity (or laxity) of the concept of "good" has been demonstrated in every era of human history, and the age of AI is unlikely to be an exception. Compounding the problem might be the open conflict, lingering disorientation, and lack of restraint that characterize many human societies today.

We wish success to our species' gigantic project, but just as we cannot count on tactical human control in the longer-term project of coevolution, we also cannot rely solely on the supposition that machines will tame themselves. Training an AI to understand us and then sitting back and hoping that it respects us is not a strategy that seems either safe or likely to succeed. Moreover, we must recognize that humans would certainly not be unified in their own approach—some treating AI as friend and others as foe, and some (given constraints on time and resources) unable to exercise a preference but simply accepting the strategy immediately available to them.

This heterogeneity suggests the likelihood of a predictable variance in levels of safety. Even as the diffusion of AIs and the lowering costs of development might accelerate AI's alignment, they might also enhance its dangers. The digital and commercial interconnectedness of today's world means that a dangerous AI, developed anywhere, would pose a threat everywhere. The disconcerting reality is that perfection in implementation entails a high standard of performance combined with an even lower tolerance for failure. Thus, discrepancies in safety regimes should be a concern to us all.

We thus urge the coordination and acceleration of humanity's disparate alignment efforts. Together, whatever the project, we will have to answer profound questions. Here are two: When the distinction between humans and machines becomes blurry, what is our minimum threshold for being treated as a species? If forced to compromise with machines, what is our collective nonnegotiable red line? Without a shared understanding of who we are, the human race risks fully relinquishing to AI the foundational task of defining our value and thereby justifying our existence.

In this light, it must be said forthrightly that, should it appear impossible to realize a regime of reliable technical strategic control, we should prefer a world with no AGI at all to a world in which even one AGI remains unaligned with human values. To be sure, how to achieve consensus—on what those human values are, how they are to be adjudged and agreed upon, and how they should be assessed, activated, and deployed—is the philosophical, diplomatic, and legal task of the century. Yet we are compelled by the exigencies of the moment and the benefits of the technology to establish and, to the maximum possible, unify moral constraints on the not-human agents that humanity is now birthing.

With sufficient democratic input as well as legal and technical expertise, with extraordinary caution, and mindful always of the misuses and malfunctions we have here described, we believe it will be possible to inculcate a moral baseline into artificially intelligent machines, and crucially to do so in lockstep with our fellow humans. Thereby we may cross the threshold to a new age with, if not utter confidence, at least informed and solemn hope.

DEFINING HUMANITY

As machines increasingly assume human qualities (and if some humans enhance themselves to take on machine-like qualities), lines will become hazy. What is AI and what is human will change and, in some cases, merge. In coming to judge how we must keep pace with AI, humans will therefore need to assert more clearly what it is that distinguishes us from machines. How, then, will we compile and compress the entire range of human experience for easy comprehension by AI?

To preclude either our demotion or our replacement by machines, some will want to lay claim to difference via our proximity to divinity. Others will wish to ascertain more tactical conclusions: which kinds of decision-making can be delegated to machines and which cannot. We propose the articulation of an attribute, or set of attributes, that most of humanity could rally behind and orient around: one that will supply a floor underneath what is preferable but not a ceiling to what might be possible.

As a starting point, we would encourage a definition of *dignity*. Without a shared definition, we will be unable to reach agreement if and when AI is being used as a method or justification for the violation or erosion of dignity, and we would thus be hamstrung in our response. Without a definition of dignity, we would not know if and when AI, given enough faculties, could become a being of dignity, could stand fully in place of a human, or could be entirely unified with a human. An AI, even if sustainably proved to be not-human, might instead constitute a member of a separate, similarly dignified category that would nonetheless deserve its own, equal standard of treatment.

One conception of dignity, developed by the eighteenth-century philosopher Immanuel Kant, is centered on the inherent worth of the human subject as an autonomous actor, capable of moral reasoning, who must not be instrumentalized as a means to an end. Could AIs come to fulfill those requirements? A definition of dignity, we believe, would help humanity answer some of these questions and encourage inclusive coexistence with AI while avoiding reckless attempts at premature coevolution.

Both to retain an understanding of ourselves and to ensure that an appropriate conception of humanity is transmitted to machines as they learn, we humans will need to recommit ourselves to more than theoretical definitional work. The actuation of agency, curiosity, and freedom, renewing and exercising our inquisitiveness about other humans, about the natural world, about the universe, and about the possibility of the divine, will assist us in our ongoing participatory redefinition of the lines of humanity.

We will especially need to ensure that, beyond conventional ideas of worth like value and power, intrinsic human *importance* becomes one of the variables that defines machine decision-making. For instance, mathematical precision may not easily encompass the concept of mercy. Even to many humans, mercy is an inexplicable ideal, if not a miracle. For its part, and without considering the rules-based aspect, a mechanical intelligence might operate by valuing machine overperformance more highly than human performance. In such a situation, could the logic behind mercy, even if it cannot be formally taught, perhaps be absorbed? Again, dignity—the kernel from which mercy blooms—might serve here as part of the rules-based assumptions, or the iterative learning, of the machine.

Clear articulation of specific defining human attributes—particularly those that are, like dignity, widely integrated into both international political instruments and global faiths—could guide human efforts during periods of disorientation, including the choice between activity and passivity, the potential limits of self-evolution, and the precise transformation of AI in the direction of the human.

To illustrate the concept's utility, take as a point of departure the following definition: Dignity is a quality that inheres in creatures who, born vulnerable and mortal and thus full of insecurity and fear, and despite their natural inclinations, can and do exercise their freedom not to follow their conception of evil but to choose their conception of good. In other words, those who can achieve dignity should, and those who do achieve it merit a special level of respect.

Undoubtedly, this definition is imperfect. It may leave out some living humans who are unable to make decisions—for instance, a person conscious but unresponsive—whom we nonetheless believe to be deserving of recognition as beings with dignity and thus entitled to respect. In this connection, perhaps the definition should be revised to indicate that dignity, once won, does not depart from us should we become unable to continue the actions that won it to begin with. A thousand such hypotheticals and emendations could be imagined.

Does this dictate our urging those who have chosen passivity in the face of powerful AI to exhibit agency and activity instead? If active, practical commitment is part of any moral ideal, then yes, the realization of dignity should be advanced. Under our definition, freedom is part of the human ideal, and thus we may expect—or even demand—that humans retain and exercise the power of conscious choice in the age of AI.

Under this definition, can AI itself possess dignity? Likely not—for AIs are not born, do not die, feel neither insecurity nor fear, and do not have natural inclinations or individuality such that conceptions of evil or good could be considered "theirs." While AIs of the near future may appear otherwise, having personalities, expressing emotions, telling jokes, and recounting personal histories, under this framework they should be treated, philosophically, like literary characters. They may embody elements of humanity; they are not real in a moral sense.

Even the greatest literary character—Shakespeare's Hamlet, for instance—is no more than a special combination of words, once written on a page and now reproduced many times over. "Hamlet" has no ability to feel a prickle in the eye, a turn of the stomach, a hot wave of frustration at thwarted expectation. "Hamlet" has no freedom to make a new choice. "Hamlet" is trapped inside his play. "Hamlet" is not a human, but a picture of a human. AI, made of strings of code and hunks of silicon, is much the same.

No doubt, some humans would therefore decry this definition of dignity as unhelpful, both philosophically and in substance. It might be criticized for being too low a common denominator—vague enough to appease all parties because of its excessive malleability—and concurrently for not capturing the idea that humans are worth preserving for their own sake and that, in some way, we are exceptional beyond our ability simply to survive. The philosopher Arthur Schopenhauer cursed dignity as "the shibboleth of all perplexed and empty-headed moralists."[14]

But dignity, as we define it, usefully holds up our fragility and our potential for failure as well as our vitality, freedom, and ability to manifest our beliefs. It points at the good that we are capable of but have not achieved, and urgently, scoldingly, whispers: *Go*.

Still, admittedly, dignity alone cannot be enough. Other attributes should be specifically considered and perhaps added to the conception of humanity as it will be deployed in our coming partnership with AI. But our ability to define and sustain core elements of humanity as a baseline for AI's comprehension of humanity at large is now a problem of existential significance, and the work to inculcate our definitions must be done now.

No definition will remain static; no doubt, as our own identities shift, we will need to evolve AI's understandings in perpetuity. Meanwhile, others more capable than ourselves will continue to advance our collective thinking on the dynamic relationship between "us" and "them," and their genius might yield a conception of humanity that more strongly (even if futilely) aims at ensuring our survival as a recognizable species. Even as they do, however, we should all strive for a definition, and a program, that goes further and elevates the human condition to new heights. For may not AI itself yet turn out to offer the strongest evidence of humanity's ability to become an active participant in creation?

OUR CHALLENGE

To guarantee tactical control by humans of every AI decision would require us to stifle the benefits of AI deployment. Relying on the substratum of human morality as a form of strategic control, while relinquishing tactical control to bigger, faster, and more complex systems, is likely—eventually, and perhaps sooner than we imagine—the way forward for AI safety. Profit-driven or ideologically driven purposeful misalignments are serious risks, as are accidental misalignments; overreliance on unscalable forms of control could significantly contribute to the development of powerful but unsafe AI. The integration of the human into the internal workings of teams of AIs, including by way of AIs to govern AIs, seems the most reliable path forward.

While the development of a humane (or humanistic) AI is our priority, we recognize some potential role for artificial humans. To the extent that we are able to develop individual self-engineering for specific abilities, thereby enabling some humans to match aspects of a future AI's intelligence, such a project could be useful. Of course, that effort would need to be the product of individual choice. Our authorial caution here is reflective of our collective dilemma: Evolution cannot merely be replaced by design, for this would be the forsaking of humanity. But ceding the project of discovery itself—whether spiritual, physical, scientific, or philosophical—would effectuate the same.

In the age of AI, the tension between the need to *design* ourselves and the need to *align* our creation may very well become our orienting compass. Both are as aspirational as they can also be conservative. It is yet to be determined to what extent the two needs will be viewed as fundamentally in contradiction with each other. If we give full play to AI's powers for unlimited exploration in this new age, we risk passivity or, worse, paralysis. But if we maximize our control, giving at least the illusion of safety, we circumscribe the extent of our fullest potential. Can we effectively mediate the exercise of our powers—our ever-advancing abilities of design and discovery—by reasserting a common and evolving conception of humanity?

We desire a future in which human intelligence and machine intelligence empower one another. To achieve that end, each intelligence must have an adequate understanding of the other. Defining who we are is just the first step, for to be human is not a constant. Far more work is needed to render our machines, and ourselves, transparent, legible, and reliable. Even if that goal is achieved for the snapshot of a moment, calibrating and sharing our truths and realities will be an intensive and ongoing project. In this way, the questions of coevolution and coexistence are not simply to be answered; they are to be enacted.

CONCLUSION

TO US, THE advent of AI is the start of an odyssey of the spirit as much as it is an expedition of logic and truth. We cannot properly consider, let alone prepare for, coexistence and coevolution with not-human life by relying on rationality alone—whether that of a machine or that of a human. Rather, something more is required—something fundamentally human.

It has long been hypothesized that our universe is something akin to an ancient game of chess—continuously played by entities older and larger than our present observable reality. The longer we have observed the movements on the board, the likelier it has become that we would eventually catch on to a few rules of play. After watching long enough, we might even begin to play ourselves. The move from passive observation to active participation is not a leap of logic. Translating principles into action is always a leap of faith.

Asked about his religious beliefs, Albert Einstein responded:

> We are in the position of a little child, entering a huge library, whose walls are covered to the ceiling with books in many different tongues. The child knows that someone must have written those books. It does not know who or how. It does not understand the languages in which they are written.
>
> The child notes a definite plan in the arrangement of the books, a mysterious order, which it does not comprehend, but only dimly suspects.
>
> That, it seems to me, is the attitude of the human mind, even the greatest and most cultured, toward God.
>
> We see a universe marvelously arranged, obeying certain laws, but we understand the laws only dimly. Our limited minds cannot grasp the mysterious force that sways the constellations.[1]

Our minds remain childlike with respect to God, our world, and now our newest creations. Understanding what we have wrought, in a logical and spiritual sense, will be an essential step toward maturity as a species. But we must also make a leap, the leap of faith, to move beyond examination and into intervention. This will require us to act in a state of uncertainty: the quagmire of human leaders since the dawn of time. Action has never meant the privilege of total control—indeed, quite the opposite. Nor will it do so in the age of AI; we do not expect to know our future destiny more than we have known it in prior epochs.

Our lack of control does not require us to abandon reason or, worse still, to give up our investment, and our inclination to act, in the real world. However, as humans enter a new cycle that involves direct partnership with AI, the particular dynamics spelled out in this volume will encourage us to lean into new and old forms of inquiry alike. Our success will depend on a commitment to defining and acting upon our moral convictions. That will require steadfast courage—and a consistent strategy—as new truths alter preexisting conceptions.

In fact, it is moral purposes that encourage our continued progress. Moreover, it is the very existence of an essential substratum of human morality that enables humanity's transcendence of the dichotomy between human control (tactical, at best) and human benefit (abundant, by any definition). It is a fundamental belief in the reality of human dignity that allows us, the authors of this book, at once to recognize the promise of human-AI alignment and to accept the necessity of faith in navigating the years ahead as science advances and reveals ever greater mysteries.

Despite our many shared human qualities, we cannot expect unity in the choices that lie ahead. What some see as an anchor to steady ourselves in the storm, others see as a leash holding us back. What some praise as necessary steps toward a pinnacle of human potential, others see as a headlong rush into an abyss.

In this case, instinctive emotional divergences—and the subjective lines that are drawn by all parties—will create an unpredictable and combustible situation. Increasingly stark positions of potential "winners" and "losers" will intensify the pressure of these circumstances. The fearful will slow their own development and sabotage that of others. The overconfident will disguise their powers and, in secret, speed up their work. The timeline of coming crises will be accelerated beyond prior human experience; quickly, we will be engulfed, and it is not clear whether or how we will survive.

Might AI cause these coming crises, and then act as our savior—manufacturing problems that only it can solve, if only to prove its necessity and to remind us of our dependency? Again we return to the dilemma that has motivated much of this volume: the excruciating choice between control and utility, between the comfort of the historically independent human and the possibilities of an entirely new partnership.

That choice is difficult and necessary. It is also solvable provided one assumes in humanity a true, definable, and intrinsic goodness. We place high confidence in the technical effort to imbue our machines with that goodness. At the same time, however, and even if our machines were provably and reliably aligned with human morality, delegating responsibility and authority to them would be a monumental decision, affecting every facet of our ability to retain our human relationships, political structures, and individual and collective identities. Nostalgia for the premier status to which our species has recently grown accustomed would be widespread. For some, no matter the eventual trajectory of AI, a world transformed—even for the better—may feel scarcely different from an abrupt ending to our bedrock reality.

A further unsolved question is *Who will decide?* Who will make the choice to delegate, or not delegate, responsibility and authority? Who will give, or withhold, resources? How will any one set of deciders communicate, converge—or come into conflict with—others attempting to make the same decisions elsewhere? Are we choosing those individuals, those fallible humans, now? Have we, unwittingly, already chosen?

We authors hope to alert these decision-makers, whoever they are, to the decisions that confront us even now and to what may lie ahead. But our aim is not baldly to instill a sense of apprehension about the rise of AI. An abrupt termination in the application of AI's powers could itself precipitate a crisis. Deceleration may be even less politically manageable than our current path, creating significant dangers for those of slower momentum and destroying the hopes of those yearning for further advancements.

Neither blind faith nor unjustified fear can form the basis of an effective strategy; one needs self-doubt to have knowledge but self-confidence to act. Indeed, in the age of AI, this is all the more urgent. We must try to understand the challenges that AI will present even as we lack the prior exposure or the essential experience to guarantee the accuracy of our comprehension. And even as we navigate this daunting task, we must also, to avoid a passive future, surmount the many difficulties already facing our species.

While some may view this moment as humanity's final act, we perceive instead a new beginning. The cycle of creation—technological, biological, sociological, political—is entering a new phase. That phase may operate under new paradigms of, among other things, logic, faith, and time. With sober optimism, may we meet its genesis.

ACKNOWLEDGMENTS

In dedicating this book to the memory of Dr. Henry A. Kissinger, we, his two coauthors, have had it in mind at once to salute his stunning accomplishments as a world-renowned statesman, to pay tribute to the sheer range and depth of his strategic thinking—in no single context more striking than in his grasp in his mid-nineties of the intricacies of artificial intelligence (AI)—and, more personally, to mark the greatness of one who was also our close mentor and friend.

In the following paragraphs, we now gratefully recognize in turn some among the many colleagues and companions who have aided us in conceiving and preparing this book's ambitious investigation of AI: a "matter," in our own urgent words, "of the utmost importance to the future of humanity."

Influencing our thinking about that matter, while simultaneously providing us with crucial information and insight into the subject's technological implications, were Demis Hassabis, Dario Amodei, Daniel Huttenlocher, Graham Allison, Mustafa Suleyman, Maithra Raghu, James Mankiya, Reid Hoffman, and Sam Altman. We are greatly in their debt.

Several key collaborators contributed to the drafting, revising, and shaping of the book's substance. Nancy Kissinger—"the inspiration of my life," as her husband justly noted in dedicating to her his penultimate book, *Leadership* (2022)—lent this project her indomitably vigilant and tender surveillance.

Eleanor Runde was foremost among professional contributors. With eloquence, learning, and vision, Eleanor translated to the page her extensive discussions over the years with Dr. Kissinger, thereby cocreating the book's essential foundation, structure, and contents. Subsequently, working in close consultation with his trusted friend and associate Neal Kozodoy, and with us coauthors, Eleanor proceeded to re-subject every chapter to her acute mastery of detail, her zealous fidelity to the author's intent in both text and context, and her subtle editorial finesse.

Joining the project at its midpoint, John Ferguson further expanded the book's arguments with energy and skill. Displaying a helpful and unusual facility with history and mythology, he also enlivened its prose. With mentorship gladly provided by one of us, and through tireless collaboration, he assumed a consequential role in the drive to see the book's manuscript through to the end.

When it comes to publishers, we happily inherited Little, Brown, the firm that a few years earlier had successfully brought out *The Age of AI*. Its executive editor, Alexander Littlefield, fortified us with aid and encouragement while displaying a welcome insistence on narrative clarity and an equally refreshing sensitivity to nuance. We were no less fortunate to enjoy the expert ministrations of Michael Noon, production editor, while also benefiting on the sidelines from the sound strategic counsel of Robert D. Blackwill and Lyndsay Howard. Our agent Andrew Wylie gave the book vital representation. Throughout, J. Paul Bremer, Dr. Kissinger's literary executor, and Joel Klein, authorized to review and consult upon decisions relating to works in progress, proved stalwart and deeply informed guardians of their dear friend's legacy.

In the final phases of our own labors as coauthors of this book, and for its marketing and promotion, invaluable support came from the team at Eric Schmidt's office — especially Janine Brady, Nathalie Bussemaker, Robert Esposito, Gabe Medina, Andrew Moore, and Selina Xu, as well as Helen Dunn, Matthew Hiltzik, and Madeleine Weast at Hiltzik Strategies.

Selflessly continuing, and extending, their own decades of devoted service to Dr. Kissinger, Theresa Amantea, Jody Williams, and Jessee LePorin have remained indispensable to the end — and beyond.

NOTES

CHAPTER 1

1. Antonio Pigafetta, *The First Voyage Round the World, by Magellan*, trans. from the accounts of Pigafetta circa 1525 (London: Hakluyt Society, 1874).
2. Ernest Shackleton, *Diary of Ernest Shackleton*, January 9, 1909.
3. Mills Leif, *Frank Wild* (Whitby: Caedmon of Whitby, 1999). Accessible at the State Library of New South Wales.
4. Colin Schultz, "Shackleton Probably Never Took Out an Ad Seeking Men for a Hazardous Journey," *Smithsonian Magazine*, September 10, 2013.
5. Cited in María Jesús Benites, "'La mucha destemplanza de la tierra': Una aproximación al relato de Maximiliano de Transilvano sobre el descubrimiento del Estrecho de Magallanes," *Orbis Tertius*, 17, no. 19 (2013).
6. Zoe Hobbs, "How many people have gone to space?" *Astronomy*, November 17, 2023 https://www.astronomy.com/space-exploration/how-many-people-have-gone-to-space.

7. See Edward L. Dreyer, *Zheng He: China and the Oceans in the Early Ming Dynasty, 1405–1433* (New York: Pearson Longman, 2007).

8. Roshdi Rashed, "A Polymath in the 10th Century," *Science*, August 2, 2002.

9. See the Shammasiyah observatory set up around 828 CE on the orders of Caliph al-Ma'mun in Baghdad under the purview of the scientific academy of the House of Wisdom in Baghdad.

10. See *The Life and Writings of Averroes*, trans. Nishikanta Chattopadhyaya (Leipzig: Cheekoty Veerunnah & Sons, 1913).

11. See works by Shen Kuo: https://www.gutenberg.org/ebooks/author/2419; see also https://ia600301.us.archive.org/24/items/pgcommunitytexts27292gut/27292-0.txt.

12. Boris Menshutkin, *Russia's Lomonosov: Chemist, Courtier, Physicist, Poet* (Princeton: Princeton University Press, 1952), 15.

13. Orrin E. Dunlap Jr, "An Inventor's Seasoned Ideas: Nikola Tesla, Pointing to 'Grievous Errors' of the Past," *New York Times*, April 8, 1934.

14. Peter Martin, "Von Neumann: Architect of the Computer Age," *Financial Times*, December 24, 1999.

15. See Edward O. Wilson, *Consilience: The Unity of Knowledge* (New York: Vintage Books, 1998), 326.

16. See Marcelo Gleiser, *The Island of Knowledge: The Limits of Science and the Search for Meaning*, 1st ed. (New York: PublicAffairs, 2015), 8: "A vast ocean surrounds the Island of Knowledge, the unexplored ocean of the unknown, hiding countless tantalizing mysteries."

17. Demis Hassabis, "AlphaGo: using machine learning to master the ancient game of Go," The Keyword Google Blog, January 27, 2016.

18. Videos of Lee Sedol vs. AlphaGo, Game 2, Move 37, are available online. For further discussion, see Cade Metz, "In Two Moves, AlphaGo and Lee Sedol Redefined the Future,"

WIRED, March 16, 2016; Graeme S. Halford et al., "How Many Variables Can Humans Process?," *Psychological Science* 16, no. 1 (January 2005): 70–76.

CHAPTER 2

1. See Richard Danzig, "Machines, Bureaucracies, and Markets as Artificial Intelligences," *Center for Security and Emerging Technology*, January 2022; Henry Farrell, Cosma Shalizi, "Artificial intelligence is a familiar-looking monster," *The Economist*, June 21, 2023; for further discussion of the printing press, see Samuel Hammond, "AI and Leviathan: Part I," *Second Best Substack*, August 23, 2023; for further reading on the analogy between AI and corporations, see a more detailed history of the East India Company.

2. For an extended discussion of AI metaphors, see Matthijs Maas, "AI is like…: A literature review of AI metaphors and why they matter for policy," *Legal Priorities Project*, October 2023.

3. Take, for instance, the lawyer who in a court case submitted a brief, written by ChatGPT, in which the model manufactured citations to false "precedents." The judge in this proceeding levied sanctions against the attorney and his colleagues. See Larry Neumeister, "Lawyers submitted bogus case law created by ChatGPT. A judge fined them $5,000," *AP News*, June 22, 2023.

4. See @porby, "Why I think strong general AI is coming soon," September 28, 2022, *LessWrong* (lesswrong.com).

5. Charles Darwin, *On the Origin of Species* (London: John Murray, 1859), 439.

6. Greg Kestin, "The Biggest Puzzle in Physics: Reconciling Quantum Mechanics and General Relativity," *PBS*, February 14, 2018.

7. Plato (380 BC), "The Allegory of the Cave" in *Plato: Collected Dialogues*, trans. P. Shorey (New York: Random House, 1963), 747–52.

8. See Christopher Olah (@ch402), "High-Low frequency detectors found in biology!...," X (Twitter), March 23, 2023, 11:52 a.m.; Geoffrey Hinton et al., "The Forward-Forward Algorithm: Some Preliminary Investigations," *arXiv*, December 27, 2022.

9. Demis Hassabis et al., "Neuroscience-Inspired Artificial Intelligence," *Neuron* 95, no. 2 (July 19, 2017): 245–58.

CHAPTER 3

1. René Descartes, "Sixth Meditation," *The Philosophical Writings of Descartes*, trans. John Cottingham et al., vol. 2 (Cambridge: Cambridge University Press, 1984), 55.

2. Alfred North Whitehead, *Process and Reality: An Essay in Cosmology,* 2nd ed. (New York: Free Press, 1979), 15.

3. See "Debate: Do Language Models Need Sensory Grounding for Meaning and Understanding?" New York University Center for Mind, Brain, and Consciousness, March 24, 2023: https://wp.nyu.edu/consciousness/do-large-language-models -need-sensory-grounding-for-meaning-and-understanding.

4. Lauren Jackson, "What If A.I. Sentience Is a Question of Degree?" *New York Times*, April 12, 2023.

5. This is true under some theories of quantum physics, in which observation creates objective change in reality. It is also true in the sense of the human understanding of a reality in which AI's observation can create *subjective* change; see Marcelo Gleiser, *The Island of Knowledge* (New York: PublicAffairs, 2015), prologue: "Our perception of what is real evolves with the instruments we use to probe Nature."

6. See Ilya Sutskever's comments in Ross Andersen, "Does Sam Altman Know What He's Creating?," *The Atlantic*, July 24, 2023.

7. For further exploration of the idea *Homo technicus*, see Henry A. Kissinger, Eric Schmidt, Daniel Huttenlocher, "ChatGPT Heralds an Intellectual Revolution," *Wall Street Journal*, February 24, 2023.

CHAPTER 4

1. See Salvador de Madariaga, *Hernán Cortés: Conqueror of Mexico* (New York: Macmillan, 1941), 99.

2. The story of the encounter between Cortés and Montezuma remains a disputed episode of history. For Spanish accounts, see Bernal Díaz del Castillo's "The True History of the Conquest of New Spain" (Historia verdadera de la conquista de la Nueva España), late sixteenth century; Hernan Cortés, "Cartas de Relación" (Letters of Relation), between 1519 and 1526; "The Florentine Codex" (Historia general de las cosas de Nueva España), late sixteenth century; Apostolic Nuncio Bernardino de Sagahun's Letter, 1524. For an alternative interpretation which questions the historical account of the Spanish, see Camilla Townsend, "Burying the White Gods: New Perspectives on the Conquest of Mexico," *The American Historical Review*, vol. 108, no. 3 (June 2003), pp. 659–87, *Oxford University Press*, the "Annals of Tlatelolco," sixteenth century, and Diego Duran and Alfredo Chavero, "Apendice-Explicacion del Codice Geroglifico de Mr. Aubin de Historia de las Indias de la Nueva España y Islas de Tierra Firme," vol. II, 1880, 71.

3. G. K. Chesterton, "Lecture 65: Christendom in Dublin," in *Collected Works*, vol. XX (San Francisco: Ignatius Press, 2002).

4. See in particular the Kālacakra.

5. See remarks by Chamath Palihapitiya at the Stanford Graduate School of Business, Nov. 13, 2017, https://www.youtube.com/watch?v=PMotykw0SIk.

6. See Alexis de Tocqueville, *Democracy in America*, trans. Henry Reeve, Esq., in two volumes (London: Saunders and Otley, 1835; New York: J. & H.G. Langley, 1840).

7. Wang Yangming, *Instructions for Practical Living or Record of Transmitting the Mind (Chuanxilu)*, posthumously compiled by his disciples based on his teachings and discussions after his death in 1529.

8. Abir Taha, "Nietzsche's Superman," *Artkos* (UK: Artkos Media, 2013), 93.

9. Al-Farabi, *Al-Farabi On the Perfect State*, trans. Richard Walzer (Oxford, Clarendon Press, 1985), 253.

10. See T.C.A. Raghavan, *Attendant Lords: Bairam Khan and Abdur Rahim* (Uttar Pradesh: HarperCollins, 2017).

11. See Niccolò Machiavelli, *The Prince*, trans. Tim Parks (London: Penguin Classics, 2009).

12. Leo Strauss, *What Is Political Philosophy?* (Chicago: University of Chicago Press, 1959).

13. See Johan Norberg, *The Capitalist Manifesto* (London: Atlantic Books, 2023).

14. Leo Tolstoy, *War and Peace*, trans. Louise and Aylmer Maude (Chicago: Encyclopedia Britannica, 1952), 646.

15. See Friedrich Hayek, "The Use of Knowledge in Society," *The American Economic Review*, September 1945, and Thomas Sowell, *Knowledge and Decisions* (New York: Basic Books, 1996), which builds further upon Hayek.

16. See Hannah Arendt, *The Origins of Totalitarianism* (New York: Harcourt, Brace, Jovanovich, 1951).

17. See Friedrich Hayek, *The Road to Serfdom* (Chicago: University of Chicago Press, 1944).

18. Simon McCarthy-Jones, "Artificial Intelligence is a totalitarian's dream—here's how to take power back," *The Conversation*, August 12, 2020, https://theconversation.com /artificial-intelligence-is-a-totalitarians-dream-heres-how-to -take-power-back-143722.

19. Ibid.; see Yuval Noah Harari, *Homo Deus: A Brief History of Tomorrow* (New York: Harper, 2017).

20. Immanuel Kant, *Kant's Principles of Politics*, trans. W. Hastie (Edinburgh: T. & T. Clark, 1891), 36.

21. Hesiod, *The Theogony* (New York: Start Publishing, 2017); Aeschylus, *Prometheus Bound* trans. Deborah H. Roberts (Indianapolis: Hackett Publishing Company, 2012).

22. Inspired by remarks made by Lawrence H. Summers at the Harvard College China Forum, April 17, 2022.

23. See Leo Tolstoy, *War and Peace*, trans. Louise and Aylmer Maude (Chicago: Encyclopedia Britannica, 1952).

CHAPTER 5

1. Paul Scharre, "America Can Win the AI Race," *Foreign Affairs*, April 4, 2023.

2. William J. Broad et al., "Israeli Test on Worm Called Crucial in Iran Nuclear Delay," *New York Times*, Jan. 15, 2011.

3. Interview with Dario Amodei, CEO of Anthropic by Dwarkesh Patel, https://www.dwarkeshpatel.com/p/dario -amodei#details.

4. Jeremy Hsu, "China's first underwater data centre is being installed," *New Scientist*, December 4, 2023.

5. Walter Pincus, "Soviets Had Chance to Develop First A-Bomb, Historian Says," *Washington Post*, July 27, 1979.

6. Graham Allison and Eric Schmidt, "Is China Beating the U.S. to AI Supremacy?" Avoiding Great Power War Project at the Harvard Kennedy School Belfer Center for Science and International Affairs, August 2020, https://www.belfercenter .org/sites/default/files/2020-08/AISupremacy.pdf.

7. Google DeepMind and Meta AI have both already built programs that have dominated humans at the game of Diplomacy: Google DeepMind: János Kramár, Tom Eccles, et al., "Negotiation and honesty in artificial intelligence methods for the board game of Diplomacy," *Nature*, December 6, 2022; Meta: Meta Fundamental AI Research Diplomacy Team (FAIR), "Human-level play in the game of Diplomacy by combining language models with strategic reasoning," *Science*, November 22, 2022; the Chinese Academy of Sciences has gone a step further, building machine-learning algorithms trained on government databases that are being used by China's diplomats for risk assessment in vetting foreign investment projects

and the prediction of events such as political upheaval or ter-
rorist attacks; Stephen Chen, "Artificial intelligence, immune
to fear or favour, is helping to make China's foreign policy,"
South China Morning Post, July 30, 2018.

8. See Herodotus, *Histories of Herodotus*, trans. Henry Cary
(New York: D. Appleton and Company, 1904).

9. See Frank McLynn, *Genghis Khan: His Conquests, His
Empire, His Legacy* (Philadelphia: Da Capo Press, 2015), 259.

10. See the section known as "Gylfaginning" (The Beginning
of Gylfi) in Snorri Sturluson, *The Prose Edda*, early 13th
century, presented as a dialogue between the mythological
figure Gylfi, who represents a human king, and the three
gods Hárr, Jafnhárr, and Þriði.

11. See Flo Read, "Nick Bostrom: Will AI lead to tyranny?"
UnHerd, November 12, 2023, https://unherd.com/2023/11
/nick-bostrom-will-ai-lead-to-tyranny.

12. Nick Bostrom, "The Vulnerable World Hypothesis," *Global
Policy*, vol. 10, no. 4, Nov. 2019.

13. Roger Crowley, *Constantinople: The Last Great Siege 1453*
(London: Faber and Faber, 2005), 91.

14. See the ancient historical and literary concept of a *phylactery*, a magical "soul-artifact" said to be used by some sorcerers to anchor their souls to the physical world and retain
their intelligence and mind in the event their material body
is destroyed. The sorcerer is unable to be permanently killed
as long as their phylactery remains intact; hence they are
typically hidden away. Nordic stories (*Boots and His Six
Brothers*) speak of men or giants who store their hearts elsewhere so that they remain immortal despite injury in battle.

15. G. K. Chesterton, *The Illustrated London News*, January
14, 1911, cited at https://www.chesterton.org/quotations/war
-and-politics.

16. Henry A. Kissinger, *Diplomacy* (New York: Simon & Schuster,
1994), 1.

17. Henry A. Kissinger, *Nuclear Weapons and Foreign Policy* (New York: Harper & Brothers, 1957), 429.

CHAPTER 6

1. Elias Lönnrot, *The Kalevala*, 1835. Compiled from old Finnish-Karelian ballads, lyrical songs, and incantations that were a part of Finnish oral tradition.
2. Hanna-Ilona Härmävaara, "The myth of the Sampo—an infinite source of fortune and greed—Hanna-Ilona Härmävaara," TED-Ed animation, September 23, 2019, https://www.youtube .com/watch?v=71fLFOjruFc.
3. *Mahabharata*, "Adi Parva" ("Book of the Beginning"), "Vana Parva" or the "Book of the Forest." Composed around 4th century BCE. Caldron of the Dagda appears in the Irish epic tale "The Second Battle of Mag Tuired," a medieval Irish text likely composed in the Middle Ages around the 11th or 12th century. The "magic mallet" originates in the folktale "Uchide-no-Kozuchi," which translates to "The Small Magic Hammer" associated with the legendary hero Urashima Taro. Recorded and compiled during the Edo period (1603–1868) or earlier.
4. See Sam Altman, "Moore's Law for Everything," March 16, 2021, https://moores.samaltman.com.
5. See the documentary film *AlphaGo—The Movie*, produced by Greg Kohs, March 13, 2020, https://www.youtube.com /watch?v=WXuK6gekU1Y.
6. Tom Simonite, "How Google Plans to Solve Artificial Intelligence," *MIT Technology Review*, March 31, 2016.
7. Arthur W. Ryder, *The Bhagavad-Gita* (Chicago: University of Chicago, 1929), 3:15.
8. Ibid., 18:41–44. See James Hijiya, The *Gita* of J. Robert Oppenheimer, *Proceedings of the American Philosophical Society*, vol. 144, no. 2, June 2000.
9. Sam Altman, "Moore's Law for Everything."

10. Ross Andersen, "Does Sam Altman Know What He's Creating?"

11. See Daron Acemoglu, *Power and Progress* (New York: PublicAffairs, 2023), for a discussion about how the wealth generated by the key technologies of the Industrial Revolution initially accrued to only a few countries and individuals. Acemoglu argues that humans were actually quite creative and inventive during the medieval period, for example, with lots of innovations across agriculture and commerce, but the prevailing vision of that period was one where you had a small elite who argued that they had a divinely endowed power, taking all the proceeds of higher productivity and putting them into monumental cathedrals (which did not meaningfully increase productivity, public health, etc.).

12. It has been suggested, for example, that unique corporate governance structures should be forced to take nonmonetary considerations into account.

13. International Telecommunication Union (ITU), "Population of global offline continues steady decline to 2.6 billion people in 2023" (Press release: Geneva, September 12, 2023). https://www.itu.int/en/mediacentre/Pages/PR-2023-09-12-universal-and-meaningful-connectivity-by-2030.aspx.

14. Jay Olson et al., "Smartphone addiction is increasing across the world: A meta-analysis of 24 countries," *Computers in Human Behavior*, 129 (2022), 107138.

15. Takes the term "Experience Machine" from Robert Nozick, *Anarchy, State, and Utopia* (Oxford, UK: Blackwell, 1974), 42.

16. Inspired by Stanford University's founding charter (https://www.stanford.edu/about/history/):

"Universities are a multiplicity of institutes, schools, laboratories, and departments that cross-fertilize each other with ideas and innovations. We explore—in the school of science—the molecular code that makes us human, and—in the schools of letters—the culture that is equally essential to our humanity. Borne from the university's commitment to the pursuit and appreciation of knowledge..."

17. Viktor Frankl, *Man's Search for Meaning* (Boston: Beacon Press, 2006), 6: "Life is not primarily a quest for pleasure, as Freud believed, or a quest for power, as Alfred Adler taught, but a quest for meaning." Found in preface by Rabbi Harold Kushner.

18. The "four arts" (*si yi*) are *qin* (a stringed instrument), *qi* (the strategy board game of Go), *shu* (calligraphy), and *hua* (Chinese painting).

19. See Erik Hoel, "Why we stopped making Einsteins," *The Intrinsic Perspective Substack*, March 16, 2022, https://www .theintrinsicperspective.com/p/why-we-stopped-making -einsteins.

20. Lev Nikolayevich (Leo) Tolstoy, *A Confession and Other Religious Writings*, trans. David Patterson (New York: W. W. Norton, 1983), 28.

CHAPTER 7

1. See Donella H. Meadows, *Thinking in Systems* (White River Junction, VT: Chelsea Green Publishing, 2008).

2. Today, the number of people who die naturally by old age is still greater than unnatural causes by ill health.

3. Antoine de Jussieu brought back coffee plants from Java to Paris. See Deligeorges et al., *Le Jardin des Plantes et le Museum National d'Histoire Naturelle* (Paris: Patrimoine, 2004), 13–15. The Jardin des Plantes was originally created in 1635 by King Louis XIII's physicians to house the royal medicinal herb garden and overseen by them for His Royal Majesty. King Louis XIII died on May 14, 1643, age 41. For the scientific expedition to the Amazon, see the French Geodesic Mission to the Equator in the eighteenth century.

4. See Sima Qian, *Records of the Grand Historian: Han Dynasty II* (New York: Columbia University Press, 1993).

5. Jack London, quoted by his literary executor, Irving Shepard, in an introduction to a 1965 collection of London's stories. Jack London, *Jack London's Tales of Adventure*, ed. Irving Shepard (Springdale, AR: Hanover House, 1956), vii.

6. Tolstoy quotes Socrates in *A Confession and Other Religious Writings*, trans. David Patterson (New York: W. W. Norton, 1983), 43: "'We move closer to the truth only to the extent that we move further from life,' says Socrates, as he prepares for death. What do we who love truth strive for in life? To be free of the body and of all the evils that result from the life of the body. If this is so, then how can we fail to rejoice when death approaches?" Socrates discusses this in Sections 62–69 of Plato's *Phaedo*, when his friends have come to see him one last time before his execution.

7. Von Neumann's religious beliefs (or lack of them) have been the subject of much discussion. Jewish by birth, he accepted Catholic baptism in 1930 in order to marry, though he did not practice the faith and some of his colleagues saw him as "completely agnostic." It therefore came as a great surprise to them that von Neumann, when dying of cancer in the hospital, sought the ministrations of a Catholic priest, Fr. Anselm Strittmatter, O.S.B., to whom he confessed and from whom he received the last sacraments of the Catholic Church. Society of Catholic Scientists, Catholic Scientist of the Past: John von Neumann, https://catholicscientists.org/scientists-of-the-past/john-von-neumann.

8. Maurice York and Rick Spaulding, *Ralph Waldo Emerson: The Infinitude of the Private Man* (2008); Robert D. Richardson, *Emerson: The Mind on Fire* (1995); Ronald Bosco and Joel Myerson, *The Selected Lectures of Ralph Waldo Emerson* (2005).

9. Ralph Waldo Emerson, *The Complete Works of Ralph Waldo Emerson* (Boston: Houghton Mifflin, 1904), vol. 4, no. 12; lecture by Ralph Waldo Emerson, "The Uses of Natural

History" for the Boston Natural History Society at the Masonic Temple in Boston, November 5, 1833. Later refined, perfected, and published in his first book, *Nature*, in 1836.

10. The *Jardin royal des plantes médicinales* was founded in 1635; the French Revolution began in 1789; the Gallery of Evolution was inaugurated in 1889. The idea of evolution (but not natural selection as its mechanism) was written about by 70 different individuals between 1748–1859, the year that Darwin published *On the Origin of Species*.

11. The term "ice age" can be misleading because ice ages are technically classified as a mix of advancing glaciers (glacials) or retreating glaciers (interglacials). Though interglacials are relatively warm, they are still classed as part of a glacial epoch. Our current period still happens to be technically classified as an ice age as we exist during an interglacial epoch right now.

12. Fyodor Dostoevsky, *The Brothers Karamazov*, trans. Constance Garnett (New York: The Modern Library, 1900), 783.

13. Depends on how you count; other estimates say that Earth has had as many as 20 mass extinctions. Some still remain debated today.

14. Lecture by Dr. David Keith at Gustavus Adolphus College Nobel Conference: "How Might Solar Geoengineering Fit into Sound Climate Policy," September 25, 2019, https://www.youtube.com/watch?v=Ia1AWdmRsMc&t=234s.

15. Calcium looping, a process first proposed in 1999 by Japanese chemists utilized in direct air capture technologies. See Shimizu, Hirama, and Hosoda, "A Twin Fluid-Bed Reactor for Removal of CO_2 from Combustion Processes," *Chemical Engineering Research and Design*, 77, no. 1, January 1999, 62–68.

16. Estimations are complex and imprecise, but most analysis on this question produces timelines that differ only by decades.

See various analyses by the U.S. Energy Information Administration, Stanford, BP, and the Energy Institute. Few reliable sources list oil and gas reserves based on current and projected production and consumption rates lasting more than 100 years, and for coal, 200 years.

17. Nemonte Nenquimo, "This is my message to the western world—your civilisation is killing life on Earth," *The Guardian*, October 12, 2020. Nenquimo is a member of the Waorani Nation from the Amazonian Region of Ecuador.

18. Winston Churchill's "We shall fight on the beaches" speech, delivered to the House of Commons on June 4, 1940: "The New World, with all its power and might, steps forth to the rescue and the liberation of the old." See *Never Give In! The Best of Winston Churchill's Speeches* (London: Pimlico, 2004), 218.

19. See Ross Andersen, "What Happens If China Makes First Contact?" *The Atlantic*, December 15, 2017.

20. Peter Ma et al., "A deep-learning search for technosignatures from 820 nearby stars," *Nature Astronomy* 7, 492–502, January 30, 2023.

21. Referring to the Kaaba, which houses a black stone that had fallen from the sky, linking heaven and earth, on display in the center of the Masjid al-Haram in Mecca, Saudi Arabia.

22. Yeshaya Elazar, "Kefitzat Haderech: What's the Message of This Rare Form of Divine Intervention?" Chizuk Shaya (blog), November 29, 2009. https://www.chizukshaya.com/2009/11/kefitzat-haderech.html.

23. Stephen Hawking and Leonard Mlodinow, *The Grand Design* (New York: Bantam, 2010), 82.

CHAPTER 8

1. For further discussion on the use of BCIs for coevolution with artificial intelligence, see Nick Bostrom, *Superintelligence: Paths, Dangers, Strategies* (Oxford: Oxford University Press, 2014), 63–67; "Brain-Computer Interfaces and AI Alignment,"

LessWrong, August 28, 2021, (lesswrong.com); Tim Urban, "Neuralink and the Brain's Magical Future," *Wait But Why*, April 20, 2017 (waitbutwhy.com); the informal mission statement of Neuralink is *"If you can't beat 'em, join 'em."* See original tweet from Elon Musk, https://twitter.com/elonmusk/status/1281121339584114691?lang=en, July 9, 2020.

2. See Charles Darwin, *On the Origin of Species* (London: Pickering & Chatto, 1992), 403.

3. See Erich Jantsch, *The Self-organizing Universe* (Oxford, UK: Pergamon Press, 1980).

4. Daniel Dennett, *From Bacteria to Bach and Back: The Evolution of Minds* (New York: W. W. Norton, 2017), 206. Dennett quotes Rogers and Ehrlich (2008), in a study of the evolution of the Polynesian canoe, quoting the French philosopher Alain ([1908] 1956) writing about fishing boats in Brittany. For further discussion, see Edward Lee, "Coevolution of human and artificial intelligence," *Berkeley Blogs*, September 18, 2017, https://news.berkeley.edu/2017/09/18/coevolution-of-human-and-artificial-intelligences.

5. Lev Nikolayevich (Leo) Tolstoy, *A Confession and Other Religious Writings*, trans. David Patterson (New York: W. W. Norton, 1983), 77.

6. Laurance Rockefeller, Henry Kissinger, et al., Prospect for America: The Rockefeller Panel Reports (New York: Doubleday, 1961), xv.

7. The fullest account is found describing the old satyr Silenus, a tutor of Dionysus in Ovid, *Metamorphoses*, 8 AD, book 2, l. 110; other accounts exist from Aristotle's *Politics* (fourth century BCE) and Alexander Polyhistor (first century BCE); Ariel Conn, "Artificial Intelligence and the King Midas Problem," December 12, 2016, https://futureoflife.org/ai/artificial-intelligence-king-midas-problem.

8. Ron Clements et al. *Aladdin*. Disney: USA, 1992, based on the folktale "Aladdin's Wonderful Lamp," shared by Syrian storyteller Hanna Diyab in 1704 and incorporated by French

translator Antoine Galland into *The Thousand and One Nights*.

9. The Center for AI Safety recently outlined a number of existential risks to humanity that might be posed by AI's development of such capabilities and goals as self-preservation. See Dan Hendrycks and Mantas Mazeika, "X-Risk Analysis for AI Research," *arXiv*, June 13, 2022.

10. See Kevin Hurler, "Chat-GPT Pretended to Be Blind and Tricked a Human into Solving a CAPTCHA," *Gizmodo*, March 16, 2023, https://gizmodo.com/gpt4-open-ai-chatbot -task-rabbit-chatgpt-1850227471. Researchers tasked a bot with getting past a ReCaptcha: a digital guardrail designed to allow only human users into particular systems. The bot hired a human on TaskRabbit, an online service that matches users with short-term task-doers (ordinarily, cleaning an apartment or walking a dog), to solve the CAPTCHA. The human who was hired, suspicious of the request, asked if the bot was in fact a robot, and if that was why it could not solve the CAPTCHA. Lying, the bot told the real human that it, the bot, was a blind human. It is unknown whether this was because the bot had no textual inputs in which a bot would answer such a question truthfully, or because the bot deduced that it would not reach its goal if it told the truth. In any case, the human receiving the bot's communications then did as the bot had asked.

11. Eliezer Yudkowsky, "Pausing AI Developments Isn't Enough. We Need to Shut It All Down." *TIME Magazine*, March 29, 2023.

12. Pierre Bourdieu, *Outline of a Theory of Practice* (Cambridge, UK: Press Syndicate of the University of Cambridge, 1977), 164.

13. Graduation speech by former Dean of Yale Law School Guido Calabresi, Yale Law School, May 22, 2023. Specific quote is a reference to 1 Corinthians 10:12.

14. Arthur Schopenhauer, *The Basis of Morality*, trans. Arthur Bullock (London: Swan Sonnenschein, 1903), 101.

CONCLUSION

1. George S. Viereck, *Glimpses of the Great* (London: Duckworth, 1930), 373.

INDEX

INDEX

artificial humans, 185–91, 210
artificial intelligence. *See* AI;
 AI models; AI systems
artificial timescale, 48
astrophysics, 178, 188
autocracies, 89–91
autonomy, 69–70, 183–84
Aztecs, 82–84

Badiou, Alain, 188, 235n4
Bairam, Khan, 94
biological coevolution, 187–88
biological engineering, 186,
 187
Bostrom, Nick, 67, 124
Bourdieu, Pierre, 196
brain
 AI compared to, 55, 56–57, 59,
  60–61
 AI models differentiated from,
  54
 autopilot functions of, 42
 correlation of size of with
  intelligence, 53
 higher reasoning and, 60
 inference and, 42, 43
 inorganic replication of, 39
 neural networks of, 40
 physical scale of, 52–54
 predictive processing of, 56
 speed of information
  processing, 41–42
 training of, 39–40, 41, 43

capitalism, 143, 145
carbon systems, 58, 170–71, 172,
 174, 233–34n16
Center for AI Safety, 236n9
ChatGPT, 33, 223n3
chess-playing programs, 32–33,
 65
Chesterton, G. K., 86, 129
China, xvii, 14–15, 21–22,
 120–21, 156, 174–75, 231n18
Chinese Academy of Sciences,
 227–28n
choice

between passivity and activity,
 206, 207, 213–14
privilege of, 157–59
Churchill, Winston, 174, 234n18
Clarke, Arthur C., xviii
climate change, 70, 123–24, 169,
 170–74
Cold War, 13, 95, 188
computers and computing
 AI distinguished from, 40, 43
 AI's redesigning of, 144
 autonomous computer
  programs and, 73
 communication among
  computers, 75
 development of, 34, 45, 52
 history of, 185–86
 language of, 54–55
 as polymathic invention, 29–30
 supercomputers, 113, 128–29
corporations, 16, 131–32, 230n12
Cortés, Hernán, 82–84, 225n2
currencies, 149–50

Darwin, Charles, 51, 187–88,
 233n10
data centers, 53–54, 113, 128–29
DeepMind laboratory, 32, 141, 163
democracies, 89–91, 143
Deng Xiaoping, 88–89
dependence, dilemma of, 122,
 135, 189
Descartes, René, 65
digital communication, 25
digital infrastructure, 128–29,
 132–33
digital technologies, 69
diplomacy
 AI and, 3, 109, 110, 111, 121–22
 existential risks and, 77
 human tradition of, 6, 121, 122
 international architecture and,
  183
 managing emergence and, 119,
  120–21, 122
 of nation-states, 134
 space race and, 14

243

Schmidt, Eric, xx, 6
science
  AI debates and, 3
  conceptions of, 52
  Earth sciences, 169–74
  evolution and, 168–69, 233n10
  extraterrestrial life and, 174–79
  humanities and, 155–56,
    230–31n16
  instruments augmenting
    sensory apparatus, 85
  medicine and, 162–69
  synthetic materials and, 172
  university introductions to,
    155–57, 230–31n16
  warfare and, 128
scientific method, 19, 44–45, 57
security
  AI and, 109–11, 122–24,
    134–35
  diplomacy and, 109, 110
  espionage and sabotage, 111–17
  geopolitical restructuring and,
    130–33
  managing emergence and,
    118–25
  military strategy and, 109, 110
  peace and power and, 133–36
  warfare and, 110, 111, 116,
    125–30
self-improving machine, 29,
  192
sensory perception, 65, 70–71
Shackleton, Ernest, 11–12, 16
Shen Kuo, 22
silicon systems, 58, 174, 190, 201,
  208
Socrates, 92, 93, 166, 232n6
Song dynasty, 22
South Pole, 11–12, 16
Soviet Union, 13–14, 95
space race, 13–14, 15
SpaceX, 15–16
statecraft, xvii, xviii, 14–15, 88,
  94, 98–99, 107, 122
strategic decision-making, 109,
  217–18

strategies
  alignment problem and, 201–4,
    211
  artificial humans and, 185–91,
    210
  challenges and, 210–11,
    217–18
  defining humanity and, 204–9,
    210
  fundamentals of, 184
  human AIs and, 191–201
Strauss, Leo, 94, 97
Stuxnet, 112–13
supercomputers, 113, 128–29
superintelligent AI, 115, 119, 120,
  122
synthetic biology, 123, 185

technological timescale, 48–49
Tesla, Nikola, 25, 26, 29
Theodosian Walls of
  Constantinople, 125–26
time
  AI timescale, 48–49, 103, 184
  human timescales, 4, 48–49,
    184
  perceptions of, 103
Tolstoy, Leo, 99, 106, 158, 166,
  168, 190, 232n6
totalitarianism, xv, 101–2
transparency, 44–45
truth, xix, 3, 5, 44–47, 76,
  195–97, 199, 202

uncertainty, xx, 84, 117, 215
unipolarity, 118
universities, 155–57, 230–31n16

value, distribution of, 149–50
values
  AI alignment with human
    values, 5, 186, 191–92, 198,
    199–200, 215–16
  interests and, 134
Voltaire, 23, 156
von Neumann, John, 26–27, 166,
  232n7

ABOUT THE AUTHORS

Henry A. Kissinger was born in Germany in 1923, served in the U.S. Army during World War II, and taught history and government for two decades at Harvard University before becoming National Security Advisor and Secretary of State in the administrations of Presidents Richard Nixon and Gerald Ford. A recipient of the Nobel Peace Prize and the Presidential Medal of Freedom, among other awards, he authored numerous influential works on statesmanship and international relations, most recently *Leadership*, and with Eric Schmidt and Daniel Huttenlocher coauthored *The Age of AI*. Until his death in November 2023, he remained in ceaseless demand as an adviser to American presidents and many other world leaders and policymakers.

ABOUT THE AUTHORS

Eric Schmidt is a technologist, entrepreneur, and philanthropist. Joining the founders of Google in 2001, he helped grow the company from a Silicon Valley startup to a global leader in technology, first as chief executive officer and chairman, and later as executive chairman and technical adviser. In 2021, he founded the Special Competitive Studies Project, a nonprofit initiative to strengthen America's long-term competitiveness in AI and technology. Most recently, he and his wife Wendy cofounded Schmidt Sciences, a nonprofit organization working to advance science and technology that deepens human understanding of the natural world and develops solutions to global issues.

Craig J. Mundie, president of Mundie & Associates, joined Microsoft in 1992 and retired in 2014 as chief research and strategy officer. He advises Microsoft on quantum computing and cybersecurity, is a director of the Institute for Systems Biology, a technology adviser to the Cleveland Clinic, and also an investor/adviser in early-stage companies involved in AI, biotech, fusion energy, and materials science. He served Presidents Clinton, Bush, and Obama on the National Security Telecommunications Advisory Council and the President's Council of Advisors on Science and Technology. Among his honors is a Doctor of Engineering degree from Rensselaer Polytechnic Institute.